## RISK MANAGEMENT SERIES

# Designing *for* Earthquakes

A MANUAL FOR ARCHITECTS

**PROVIDING PROTECTION TO PEOPLE AND BUILDINGS**

## BACKGROUND AND PURPOSE

In 1978, the National Science Foundation supported the American Institute of Architects in the preparation of a document entitled "Designing for Earthquakes." This document, which has long been out of print, was a compendium of papers presented at the 1978 Summer Seismic Institutes for Architectural Faculty, held at the University of Illinois and Stanford University.

FEMA has long fostered a strong relationship with the architectural community. It was decided that Designing for Earthquakes, that had remained for many years a major reference for architects and related professions, should to be updated to reflect advances in technology and understanding that had occurred since the original document was published

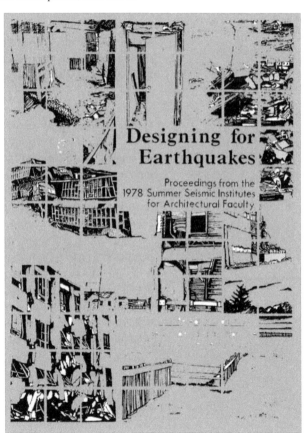

The need for updating this publication was prompted by the fact that literature on natural hazard mitigation directed towards the architectural profession is scarce in spite of the fact that architects can make a significant contribution to hazard risk reduction. While many textbooks exist on the design of structures and the nature of earthquakes, they are of a specialist nature directed to their own disciplines, and written in their own special language.

Currently no single publication exists that provides up-to-date information necessary to architects presented in a form that is attractive, readable and intelligible to a non-specialist audience. This revised publication will fill that gap.

The present publication, under the same title as the original document, is a completely new work. It follows the general approach of the original in that it consists of a series of chapters that provide the foundation for an understanding of seismic design, each authored by an expert in the field. The authors were given freedom to decide the scope of their chapters and thus this publication represents expert opinion rather than consensus. An outside expert review panel has reviewed two drafts of the publication to ensure that the selected topics are covered in an accurate, interesting and useful way.

Designing for Earthquakes: a Manual for Architects is intended to explain the principles of seismic design for those without a technical background in engineering and seismology. The primary intended audience is that of architects, and includes practicing architects, architectural students and faculty in architectural schools who teach structures and seismic design. For this reason the text and graphics are focused on those aspects of seismic design that are important for the architect to know.

Earthquakes in the United States are regional in their occurrence and while California is famous for its earthquake other states, such as Texas, have much less concern for the threat of temblors. However, architectural practice is becoming increasingly national and global, and the architect in Texas may find that the next project is in California. Thus it has become necessary for the professional architect to have some knowledge of the earthquake problem and how design seeks to control it.

Because of its non-technical approach this publication will also be useful to anyone who has an interest and concern for the seismic protection of buildings, including facility managers, building owners and tenants, building committee participants, emergency service personnel and building officials. Engineers and engineering students will also gain from this discussion of seismic design from an architectural viewpoint.

The principles discussed are applicable to a wide range of building types, both new and existing. The focus is on buildings that are designed by a team that includes architects, engineers and other consultants.

## ACKNOWLEDGMENTS

This publication has been produced by the Earthquake Engineering Research Institute (EERI) of Oakland, California, under a grant from FEMA/DHS

### Writing Team

| | |
|---|---|
| Christopher Arnold, FAIA, RIBA | Building Systems Development Inc Palo Alto, CA |
| Bruce Bolt (deceased) | Professor Emeritus, Dept. of Civil and Environmental Engineering and Earth and Planetary Science, University of California, Berkeley, CA |
| Douglas Dreger | Associate Professor, Dept. of Earth and Planetary Science University of California, Berkeley, CA |
| Eric Elsesser | Structural Engineer Forell/Elsesser Engineers Inc San Francisco, CA |
| Richard Eisner, FAIA | Regional Administrator Governor's Office of Emergency Services Oakland, CA |
| William Holmes, | Structural Engineer Rutherford & Chekene, Consulting Engineers, Oakland, CA |
| Gary McGavin | Gary L. McGavin, AIA Architecture, Professor, Dept. of Architecture, California State University, Pomona |
| Christine Theodoropoulos, AIA, PE | Head, Dept. of Architecture University of Oregon, Eugene |

### Project Team

| | |
|---|---|
| Susan Tubbesing, Project Director | Executive Director Earthquake Engineering Research Institute, Oakland, CA |
| Christopher Arnold, FAIA, RIBA Co-Project Director and Editor | President Building Systems Development Inc Palo Alto, CA |

James Godfrey
Project Coordinator

Special Projects Manager
Earthquake Engineering Research Institute,
Oakland, CA

Tony Alexander, AIA, RIBA
Publication Design and Graphics

Graphics Consultant
Palo Alto, CA

Wanda Rizer
RMS Publications, Format

design4impact
Abbottstown, PA

**External Review Panel**

Leo E. Argiris

Principal
Arup , New York, NY

Charles Bloszies, AIA

Charles Bloszies, Architecture/Structures,
San Francisco, CA

Gary Chock

President, Martin & Chock Inc.
Honolulu, HI

Charles Davis, FAIA

EHDD Architects, San Francisco, CA

Deane Evans, FAIA

Executive Director, Center for Architecture and
Building Science Research,
New Jersey Institute of Technology, Newark, NJ

Kirk Martini

Professor, Dept. of Architecture
University of Virginia,
Charlottesville, VA

Jack Paddon, AIA

Williams + Paddon,
Architects+ Planners Inc
Roseville, CA

Todd Perbix, PE

Perbix Bykonen Inc, Seattle, WA

**Project Officer**

Milagros Kennett

Architect/Project Officer,
Risk Management Series, Mitigation Division,
Building Science and Technology,
Department of Homeland Security/ FEMA

All unattributed photos are by the authors. Every effort has been made to discover the source of photos, but in some instances this was unsuccessful.

# TABLE OF CONTENTS

## CHAPTER 3 - SITE EVALUATION AND SELECTION
### Richard Eisner

## CHAPTER 4 – EARTHQUAKE EFFECTS ON BUILDINGS
### Christopher Arnold

## CHAPTER 5 – SEISMIC ISSUES IN ARCHITECTURAL DESIGN
### Christopher Arnold

## CHAPTER 6 – THE REGULATION OF SEISMIC DESIGN
### Christine Theodoropoulos

## CHAPTER 7 - SEISMIC DESIGN: PAST, PRESENT AND FUTURE
### Eric Elsesser

## CHAPTER 8 – EXISTING BUILDINGS: EVALUATION & RETROFIT
### William Holmes

## CHAPTER 9 – NONSTRUCTURAL DESIGN PHILOSOPHY
### Gary McGavin

**CHAPTER 10 – DESIGN FOR EXTREME HAZARDS**
**Christopher Arnold**

## TABLES

## FIGURES

## CHAPTER 4

## CHAPTER 5

## CHAPTER 8

## CHAPTER 9

By Christopher Arnold

## 1.1 THE BACKGROUND

Earthquakes have long been feared as one of nature's most terrifying phenomena. Early in human history, the sudden shaking of the earth and the death and destruction that resulted were seen as mysterious and uncontrollable.

We now understand the origin of earthquakes and know that they must be accepted as a natural environmental process, one of the periodic adjustments that the earth makes in its evolution. Arriving without warning, the earthquake can, in a few seconds, create a level of death and destruction that can only be equalled by the most extreme weapons of war. This uncertainty, combined with the terrifying sensation of earth movement, creates our fundamental fear of earthquakes.

The Tangshan, China, earthquake of 1976 is officially reported to have caused 255,000 deaths: foreign observers say the total may be much more. The city of Tangshan was essentially leveled as if struck by an atomic bomb (Figure 1-1).

Figure 1-1

The city of Tangshan, China, after the 1976 earthquake. The city was leveled and over 250,000 of the city's 750,000 inhabitants were killed.

SOURCE: CHINA ACADEMIC PUBLISHERS, THE MAMMOTH TANGSHAN EARTHQUAKE OF 1976, BEIJING, 1986

However, Tangshan was a city of largely nonengineered, unreinforced masonry buildings: this level of destruction is not expected in a city built in accordance with recent seismic codes.

As described in this publication, many characteristics of the site, the earthquake and the structure influence seismic performance. It is common for a group of engineered buildings to demonstrate extremely varied damage patterns within a small area that receives essentially the same ground motion. The effect of poor soils, clearly shown in the San Francisco earthquake of 1906, was demonstrated again in the Loma Prieta earthquake of 1989. The Marina district, which was built partially on fill recovered from the debris of the 1906 earthquake, suffered substantial damage, and some buildings collapsed because of the amplified

Figure 1-2

Collapsed apartment house in the Marina District of San Francisco, caused by a combination of amplified ground motion and a soft story.

SOURCE: NIST

ground motion at the site (Figure 1-2). The United States does not rate very high in deadly earthquakes compared to other countries. In the entire history of the United States, the estimated number of earthquake-related deaths is only about 3,100, of which some 2,000 are accounted for by the 1906 San Francisco quake.

The Northridge (Los Angeles) earthquake of 1994 is the most recent large earthquake in the United States. It was responsible for only 57 deaths (of which 19 were heart attacks deemed earthquake-related).

Figure 1-3

Damage in Kobe after the 1995 earthquake. Extensively damaged by air raids in World War II, Kobe was a relatively new city. Major development took place during the boom years of the 1970s and 1980s. Over 5,000 people were killed.

This was the result of the excellence of California design and construction, the time the earthquake occurred (4:31am), and because most of the earthquake's energy was directed north into a sparsely populated mountain area. However, the economic losses were estimated at $46 billion, and the earthquake was the most costly disaster in the nation's history until the recent Gulf area hurricane and floods.

However, the Kobe earthquake of 1995 showed what an earthquake centered on the downtown region of a modern city could do, even though Japan vies with the United States in the excellence of its seismic design and research. Over 5,000 deaths occurred, the majority of which happened in old timber frame buildings that had not been engineered. The earthquake sought out a weakness in the building inventory that had been overlooked. The regional economy, centered on the port of Kobe, was crippled, and large sections of the city's freeways collapsed (Figure 1-3).

At the regional and national levels, economic losses can be very high in industrialized countries for earthquakes that kill relatively few people. Even moderate earthquakes cause huge economic losses, largely due to the fragility of modern buildings' interiors, systems and enclosures.

While the low loss of life in United States earthquakes has been a cause of cautious optimism - now tempered by the experience of Kobe - increasing economic losses as a result of earthquakes are becoming a major concern. For example, in the past 30 years, earthquake losses in California, by far the most earthquake-prone state, have increased dramatically.

TABLE 1-1: Recent California Earthquakes

| Earthquake | Date | Richter Magnitude | Total Loss ($ million) |
|---|---|---|---|
| San Fernando (Los Angeles) | 2/9/1971 | 6.7 | 2,240 |
| Imperial Valley (Mexican border) | 10/15/1979 | 6.5 | 70 |
| Coalinga (Central California) | 5/2/1983 | 6.4 | 18 |
| Loma Prieta (San Francisco) | 10/17/1989 | 7.0 | 8,000 |
| Northridge (Los Angeles) | 1/17/1994 | 6.7 | 46,000 |

Table 1-1 shows a tabulation by the Federal Emergency Management Agency of earthquake losses in California between 1964 and 1994 (FEMA 1997).

Although earthquakes cannot be prevented, modern science and engineering provide tools that, if properly used, can greatly reduce their impacts. Science can now identify, with considerable accuracy, where earthquakes are likely to occur and what forces they will generate. Good seismic engineering can provide structures that can survive to a useful degree of predictability.

## 1.2 THE ARCHITECT'S ROLE IN SEISMIC DESIGN

The key figures in ensuring safe seismic design are the seismologist and the structural engineer. However, the architect initiates the building design and determines a number of issues relating to its configuration that have a major influence on the building's seismic performance. Configuration is defined as the building's size and three-dimensional shape, the form and location of the structural elements, and the nature and location of nonstructural components that may affect seismic performance. Many experienced earthquake engineers say that the architect plays the key role in ensuring the satisfactory seismic performance of a building.

To develop an effective seismic design, the architect and engineer must work together from the inception of the project so that seismic issues and architectural requirements can be considered and matched at every stage of the design process. For this process to be successful, the architect and engineer must have mutual understanding of the basic principles of their disciplines. Hence, the architect should have a basic understanding of the principles of seismic design so that they will influence the initial design concepts, enabling the engineer and architect to work together in a meaningful way, using a language that both understand. In turn, the engineer must understand and respect the functional and aesthetic context within which the architect works. The purpose of this publication is to provide the foundation for these understandings and to make the engineering and seismological language of seismic design clear to the architect and others who form the design team.

It is not intended that the study of this publication can turn the architect into a seismic engineer, capable of performing seismic analysis and creating the engineering design for the building. The intent is to help architects and engineers become better partners, not to further their separation, and to encourage a new level of architect and engineer collaboration.

Inspection and analysis of earthquake-damaged buildings play important roles in understanding the effectiveness of seismic design and construction. Although earthquake damage often appears random (one building may survive while its immediate neighbor will collapse), there are, in fact, patterns of damage that relate to the characteristics of the site discussed in Chapter 2 and Chapter 3 and to the building characteristics discussed on Chapters 4, 5, and 7.

## 1.3 THE CONTENTS OF THIS PUBLICATION

**Chapter 1** provides an introduction to some of the key issues involved in seismic design, including a summary of the effects of earthquakes worldwide and in the United States.

The nature of earthquake damage is shown graphically, to provide a context for the chapters that follow.

**Chapter 2** outlines the characteristics of earthquakes that are important for building design and discusses the nature of seismic hazard and how it is expressed. The chapter includes up-to-date information on new topics such as near-field activity and directivity.

**Chapter 3** discusses the selection and assessment of sites in earthquake hazard areas. Important collateral issues such as earthquake-induced landslide and liquefaction are covered with special attention to tsunamis.

**Chapter 4** explains the basic ways in which earthquake-induced ground motion affects buildings. This includes the ways in which buildings respond to ground motion and the characteristics of buildings that may amplify or reduce the ground motion that they experience.

**Chapter 5** explains the ways in which fundamental architectural design decisions influence building seismic performance, and shows how the building becomes more prone to failure and less predictable as the building becomes more complex in its overall configuration and detailed execution. A discussion of the ways in which architectural configurations are created leads to some speculation on the future of architectural design in relation to the seismic problem.

**Chapter 6** provides a sketch of the recent history of seismic codes as a means of ensuring a minimum level of building safety against earthquakes, and discusses some of the key concepts in seismic codes, using the *International Building Code* as a basis. The concept of performanc-based design is outlined as a means of redressing some of the flaws of current prescriptive methods of building that have been revealed in recent earthquakes.

The principles behind failures caused by architectural decisions are discussed in Chapter 4, and specific types of failure are categorized in Chapter 5. These two chapters present the core concepts with which the architect should be familiar, and make the central argument for the importance of architectural design decisions in determining a building's seismic performance.

**Chapter 7** uses a largely historical approach to show the development of earthquake resistant-design in the twentieth century. By tracing the evolution of design in the San Francisco Bay region the chapter shows the great inventiveness of earthquake engineers throughout the first half of the century, the gradual introduction of advanced methods during the latter half of the century, and the application of advanced research in base isolation methods and energy dissipation devices that has marked the last two decades.

**Chapter 8** tackles perhaps the most difficult problem facing the seismic design community, that of improving the safety of our existing seismically hazardous buildings. The chapter sketches the main issues of the existing building problem and outlines current methods of dealing with them. A common typography of building types is illustrated together with their seismic deficiencies and common retrofit techniques.

> This chapter stresses that the structural systems that are in common use have different performance characteristics, and the system selection must be properly matched to the site conditions, the architectural configuration, and the nature of the nonstructural components and systems in order to achieve the desired performance. The performance characteristics of commonly used structural systems, both those that are obsolete but still present in older buildings and those currently defined in the seismic codes, are outlined in Chapter 7, Figures 7.11A and 7.11b, and also in Chapter 8, Table 8.3.

**Chapter 9** outlines the scope of the nonstructural design problem: the protection of the components and systems that transform a bare structure into a functioning building. The chapter suggests that the nonstructural problem demands a systems approach to its solution in which the critical linkages between systems are protected in addition to the components and systems themselves.

**Chapter 10** recognizes that seismic design does not exist in a vacuum but the building must also be protected against other hazards, natural and man-made. In this regard, one issue is the extent to which protection from one hazard reinforces or conflicts with protection from another. This chapter uses a matrix to compare seismic protection methods to those of the key natural hazards: flood and high winds, the traditional hazard of fire, and the new hazard of physical attack.

## 1.4 THE BOTTOM LINE

This publication is an introduction to its subject, and deals more with principles than with the many detailed tasks that go into ensuring the seismic safety of a building. These tasks require a team approach in which all the participants in the building design and construction process must participate in a timely manner. Understanding the principles discussed in this publication will assist the design team as they search for affordable solutions that will provide building safety without compromising building function, amenity and delight.

In the confines of a document that contains a huge scope, the authors must necessarily be very selective. Seismic hazard is now clearly recognized as a national problem, and analytical and experimental research is being pursued in a number of regional centers and universities. However, there are great regional variations in seismic hazard levels. California, in particular, has had extensive experience with damaging earthquakes that have significantly influenced building design. Seismic codes, design practices and related land use and rehabilitation provisions originated in California and have been refined there for decades. Most of the material in this publication, developed by authors with first-hand experience, draws on that readily available wealth of knowledge and lessons learned.

Each chapter includes references to other readily available publications and other sources that will enable the interested reader to dig deeper into the subject matter.

# NATURE OF EARTHQUAKES AND SEISMIC HAZARDS 2

by Bruce A. Bolt and Douglas Dreger

## 2.1 INTRODUCTION

Seismology has long contributed to engineering and architecture. The founders of seismology, defined as the scientific study of earthquakes, were Robert Mallet [1810-1881], a civil engineer, and John Milne [1850-1913], a mining engineer. They were first stimulated by their field studies of great earthquakes, and then posed some basic questions, such as "What is the mechanical explanation for the damage (or lack of it) when structures are subject to seismic strong ground motion?" and "What are the essential characteristics of seismic waves that affect different structures?"

Robert Mallet, after the great Neapolitan earthquake of 1857 in southern Italy, endeavored to explain "the masses of dislocated stone and mortar" that he observed in terms of mechanical principles and the building type and design. In doing so, he established much basic vocabulary, such as **seismology**, **hypocenter** (often called the earthquake **focus**), and **isoseismal** (contours of equal seismic intensity). These nineteenth century links between seismology, engineering, and architecture have continued ever since.

A later well-known architectural example is Frank Lloyd Wright's design of the Imperial Hotel in Tokyo (Figure 2-1).

Figure 2-1

Imperial Hotel, Tokyo

SOURCE: FRANK LLOYD
WRIGHT FOUNDATION

During the planning of his ornate edifice, Wright felt many earthquakes and noted that "the terror of temblors never left me as I was planning the building." He knew that the site of the hotel would be exceptionally dangerous in an earthquake because eight feet of topsoil overlaying 60 feet of soft mud would not offer firm support. To meet this threat, he introduced a number of innovations, including shallow foundations on broad footings, supported by small groups of concrete pilings along the foundation wall. Rather than unreinforced brick walls, the building had double-course walls composed of two outer layers of brick bonded in the middle, with a core of reinforcing bars set in concrete. He designed the first floor walls to be rigid and thick; the walls of higher floors tapered upwards and contained fewer windows. He topped the structure with a hand-worked green copper roof.

Wright was also among the first architects to appreciate that the mechanical systems in buildings, such as plumbing and wiring, could be hazards in earthquakes. To lessen this risk, he ran the hotel pipes and wires through trenches or hung them from the structure so that "any disturbance might flex and rattle but not break the pipes and wiring." He also conceived the beautiful reflecting pool at the front of hotel as a reservoir of water for fire fighting.

Less than nine months after the opening of the Imperial Hotel, the Great 1923 Kanto earthquake caused enormous devastation in the Tokyo area, shattering over 5,000 buildings and creating a firestorm. The merit of Wright's reflecting pool became clear. The Imperial Hotel still stood after its battering in the earthquake, although the damage and cracking within the building was considerable.

Nowadays, seismologists can offer the architect and engineer more reliable quantitative knowledge than in 1923 concerning the earthquake hazard at a particular site, and also the patterns and intensities of the earthquake waves that are likely to shake the structure. To a large extent this is due to recent availability of more instrumental recordings of intense seismic wave motions in various geological conditions, especially near to their fault sources.

The aim of this chapter is to provide some of the latest knowledge about earthquakes that may be most relevant to architectural design. The intent is that the description should serve architects when they discuss with their clients the appropriateness of certain designs, in relation to a seismic hazard. Toward this goal the discussion covers faulting (the main cause of earthquakes) an explanation of the types of waves generated by the fault rupture, the effect of soils on the strong ground motions, and contemporary methods of estimating earthquake risk.

References are also provided to a number of research papers and books for the architect who wants to pursue the subject more deeply. Several relevant addresses of web pages on earthquakes, of which there is a diverse and growing number, are also included.

## 2.2 OBSERVATIONS OF EARTHQUAKES

### 2.2.1 Plate Tectonics and Seismicity

A coherent global explanation of the occurrence of the majority of earthquakes is provided by the geological model known as Plate Tectonics. The basic concept is that the Earth's outermost part (called the lithosphere) consists of several large and fairly stable rock slabs called plates. The ten largest plates are mapped in Figure 2-2. Each plate extends to a depth of about 100-200 km and includes the Earth's outermost rigid rocky layer, called the crust.

The moving tectonic plates of the Earth's surface also provide an explanation of the various mechanisms of most significant earthquakes. Straining and fracturing of the regional crustal rocks result from collisions between adjacent lithospheric plates, from destruction of rocky slab-like plate as it descends or **subducts** into a dipping zone beneath island arcs, and from spreading out of the crust along mid-oceanic ridges. In the United States, the most significant subduction zone is the Cascadia Zone in western Washington state, where the Juan de Fuca Plate slides (or subducts) under the America Plate (Figure 2-2). Research indicates that ruptures along this zone have resulted in very large magnitude earthquakes about every 500-600 years . The 1964 Alaska earthquake was in a subduction zone and was responsible for the greatest recorded United States earthquake. The earthquakes in these tectonically active boundary regions are called **interplate earthquakes**. The very hazardous shallow earthquakes of Chile, Peru, the eastern Caribbean, Central America, Southern Mexico, California, Southern Alaska, the Aleutians the Kuriles, Japan, Taiwan, the Philippines, Indonesia, New

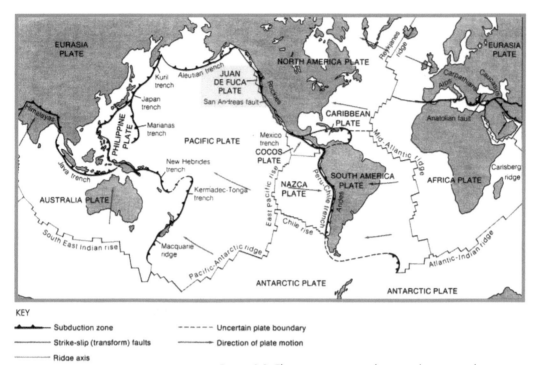

KEY

▲▲▲— Subduction zone          - - - - - Uncertain plate boundary

————— Strike-slip (transform) faults          ———⟶ Direction of plate motion

————— Ridge axis

Figure 2-2: The major tectonic plates, midoceanic ridges, trenches and transform faults.

SOURCE: BRUCE A. BOLT, *NUCLEAR EXPLOSIONS AND EARTHQUAKES: THE PARTED VEIL* (SAN FRANCISCO: W. H. FREEMAN AND COMPANY. COPYRIGHT 1976

Zealand, the Alpine-Caucasian-Himalayan belt are of plate-edge type. Earthquakes generated at depths down to 700 km also occur along plate edges by a mechanism yet unclear.

As the mechanics of the lithospheric plates have become better understood, long-term predictions of the place and size of interplate earthquakes become possible. For example, many plates spread toward the subduction zones at long-term geologic rates of from 2 to 5 cm (about one to two inches) per year. Therefore, in active arcs like the Aleutian and Japanese islands and subduction zones like Chile and western Mexico, the history of large earthquake occurrence can identify areas that currently lag in earthquake activity.

There is a type of large earthquake that is produced by slip along faults connecting the ends of offsets in the spreading oceanic ridges and the ends of island arcs or arc-ridge chains (see Figure 2-2). In these regions, plates slide past each other along what are called **strike–slip**, or **transform faults**. Considerable work has been done on the estimation of strong ground motion parameters for the design of critical structures in earthquake-prone countries with either transform faults or ocean-plate subduction tectonics, such as Japan, Alaska, Chile, Mexico, and the United States. Similar hazard studies have been published for the Himalaya, the Zagros (Iran), and Alpine regions all examples of mountain ranges formed by **continent-to-continent collisions**. Such collision zones are regions where very damaging earthquakes sometimes occur.

While simple plate-tectonic theory provides a general understanding of earthquakes and volcanoes, it does not explain all seismicity in detail, for within continental regions, away from boundaries, there are also large devastating earthquakes. These **intraplate** earthquakes can be found on nearly every continent (Yeats et al., 1997). The disastrous Bhuj (M = 7.7) earthquake in northeast India in the seismically active Kutch province was a recent example of such an intraplate earthquake (see Section 2.3.3 for an explanation of earthquake magnitude (M). In the United States, the most famous intraplate earthquakes occurred in 1811-1812 in the New Madrid area of Missouri, along the Mississippi River; another is the-damaging 1886 Charleston, South Carolina, earthquake. The Nisqually earthquake of 2001 that took place in Washington was a deep focus earthquake with a moment magnitude of 6.8. However, because of its depth of focus (32 miles), structural damage to buildings was not wide-spread and modern buildings and those recently upgraded performed well.

Shallow-focus earthquakes (focus depth less than 70 km) wreak the most devastation, and they contribute about three-quarters of the total energy released in earthquakes throughout the world. In California, for example, all of the known damaging earthquakes to date have been shallow-focus. In fact, it has been shown that the great majority of earthquakes occurring in California originate from foci in the upper ten kilometers of the Earth's crust, and only a few are as deep as 15-20 km, excepting those associated with subduction north of Cape Mendocino.

All types of tectonic earthquakes defined above are caused by the sudden release of elastic energy when a fault ruptures; i.e. opposite sides rapidly slip in opposite directions. This slip does work in the form of heat and wave radiation and allows the rock to rebound to a position of less strain.

Most moderate to large shallow earthquakes are followed, in the ensuing hours and even in the next several months, by numerous, usually smaller, earthquakes in the same vicinity. These earthquakes are called **aftershocks**, and large earthquakes are sometimes followed by very large numbers of them. The great Rat Island earthquake caused by subduction under the Aleutian Islands on 4 February 1965 was, within the next 24 days, followed by more than 750 aftershocks large enough to be recorded by distant seismographs. Aftershocks are sometimes energetic enough to cause additional damage to already weakened structures. This happened, for example, a week after the Northridge earthquake of 17 January 1994 in the San Fernando Valley, when some weakened structures sustained additional cracking from magnitude 5.5-6.0 aftershocks. A few earthquakes are preceded by smaller **foreshocks** from the source area, and it has been suggested that these can be used to predict the main shock, but attempts along this line have not proven statistically successful.

Volcanoes and earthquakes often occur together along the margins of plates around the world that are shown in Figure 2-2. Like earthquakes, there are also intraplate volcanic regions, such as the Hawaiian volcanoes in which earthquakes and volcanic activity are clearly physically related.

### 2.2.2 Earthquake Fault Types

The mechanical aspects of geological faults are the key factors in understanding the generation of strong seismic motions and modeling their different characteristics. Some knowledge of the fault type to be encountered at a site is useful to the architect because of the different types and intensities of motion that each fault type may generate.

First, the geometry of fault-slip is important (see Figure 2-3). The **dip** of a fault is the angle that the fault surface makes with a horizontal plane, and the **strike** is the direction of the fault line exposed or projected at the ground surface relative to the north. A **strike-slip** or **transform** fault involves displacements of rock laterally, parallel to the strike. If, when we stand on one side of a fault and see that the motion on the other side is from left to right, the fault is **right-lateral** strike-slip. If the motion on the other side of the fault is from right to left, the fault is termed a left-lateral strike slip. Events of strike-slip type include the 1857 and 1906 San

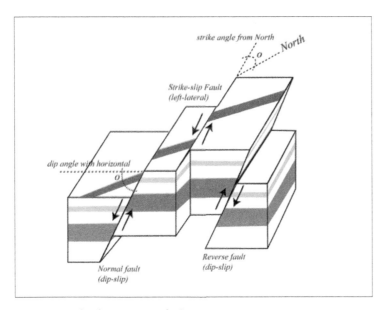

Figure 2-3: The three primary fault types.

The strike is the angle the surface trace of the fault makes with respect to geographic north. The dip is the angle the fault plane makes in the vertical with respect to the horizintal.

SOURCE: Bruce A. Bolt, *Earthquakes*, 2003

Andreas fault, California, earthquakes and more recently the 1996 Kobe, Japan ($M_W$ = 6.9), 1999 Izmit, Turkey ($M_W$ =7.6, Figure 2-4), and 2002 Denali, Alaska ($M_W$ =7.9), earthquakes.

The **right-lateral** displacement of the North Anatolian fault in Turkey from the 1999 event is shown in Figure 2-4. Catastrophic damage to multi-story buildings both near and across the fault resulted from the fault motions. A lone standing building in the foreground demonstrates that variation in building construction is also a factor in the survivability of a structure.

A **dip-slip** fault is one in which the motion is largely parallel to the dip of the fault and thus has vertical components of displacement. There are two types of dip-slip faults: the normal and the **reverse** fault.

Figure 2-4: Izmit, Turkey, 1999.

The right-lateral strike-slip fault motion (depicted by white arrows and evidenced by the offset masonry wall) pass through a collapsed structure. Note that collapsed and standing structures adjacent to the fault demonstrate both the severity of ground shaking and variation in the quality of construction.

A normal fault is one of dip-slip type in which the rock above the inclined fault surface moves downward relative to the underlying crust. Faults with almost vertical slip are also included in this category. The Borah Peak ($M_W$ = 7.3) earthquake in Idaho in 1983 is an example of a normal-type event that produced a scarp six feet high.

In a reverse fault, the crust above the inclined fault surface moves upward relative to the block below the fault. **Thrust** faults belong to this category but are generally restricted to cases when the dip angle is small. In **blind thrust faults**, the slip surface does not penetrate to the ground surface (for example, in the 1994 Northridge earthquake).

For the common shallow crustal earthquakes, seismic ground motions differ systematically when generated by strike-slip, thrust, or normal mechanisms. Given the same earthquake magnitude, distance to the site, and site condition, the ground motions from thrust earthquakes tend to be (about 20-30 percent) larger than the ground motions from strike-slip earthquakes, and the ground motions from normal faulting earthquakes tend to be smaller (about 20 percent) than the ground motions from strike-slip earthquakes. For subduction earthquakes such as the 1964 Alaska ($M_W$ = 9.2) event, the ground motions systematically differ from those generated by interface or intra-plate earthquakes. Again, for the same magnitude, distance, and site condition, the ground motions from intra-plate earthquakes tend to be about 40 percent larger than the ground motions from inter-plate earthquakes.

Reverse-fault slips have the greatest range of size, because they can grow both in the strike and dip directions. In subduction zones, the largest reverse events occur in the depth range from 0-100 km, with lengths on the order of 1,000 km. The 1960 Chile and 1964 Alaska mega-earthquakes ($M_W$ = 9.5 and $M_W$ = 9.2, respectively) are examples of this type. The 1994 Northridge, California, earthquake, despite its moderate size ($M_W$ = 6.7), inflicted considerable damage and casualties because of its location on a blind thrust beneath a heavily populated region. In most cases however, fault slip is a mixture of strike-slip and dip-slip and is called **oblique** faulting, such as occurred in the 1989 Loma Prieta ($M_W$ = 6.9) earthquake in central California. In the latter case also, the fault slip was not visible at the surface of the ground but was inferred from seismological recordings. Large scale thrusting of the ground surface was very evident along the Chelungpu fault in the 1999 Chi Chi earthquake ($M_W$ = 7.6) in Taiwan (see Figure 2-5).

It is at once obvious that any description of seismicity requires a measure of earthquake size, for comparison between earthquakes and between seismic hazard zones. As in classical mechanics, a suitable quantity to characterize the mechanical work done by the fault rupture that generates the seismic waves is the mechanical moment. In these terms we can

Figure 2-5: This building near Juahan, in Taiwan, was lifted several feet by the fault. Fault rupture runs just near the side of the building, down the alley. The white lines highlight the offset ground surface. There was no apparent damage to the building.

consider the seismic moment that is, as might be expected, proportional to the area of fault slip A multiplied by the slip distance D.

**Fault offset** a poses high risk for certain types of structures. When such structures, including dams and embankments, must be built across active faults, the design usually incorporates joints or flexible sections in the fault zone. The maximum horizontal offset in the 1906 San Francisco earthquake was about 18 feet.

## 2.2.3 Earthquake Effects

There are many earthquake effects related to the geology and form of the earth that are of significance for architects. In the most intensely damaged regions, the effects of severe earthquakes are usually complicated. The most drastic effects occur chiefly near he causative fault, where there is often appreciable ground displacement as well as strong ground shaking (e.g. Figure 2-4); at greater distance, noticeable earthquake effects often depend on the topography and nature of the soils, and are often more severe in soft alluvium and unconsolidated sediment basins. Some remarkable effects are produced in bodies of water such as lakes, reservoirs, and the sea.

● Ground Shaking Intensity

Efforts to measure the size of an earthquake by rating microseismic data in the affected area go back to the 19th century. Before the invention of instrumentally based seismic magnitude, the most common historical scale rated the relative "intensity" of an earthquake. This measure is not capable of strict quantitative definition because seismic intensity at a particular point of the Earth's surface depends on many factors, including the source moment M0, area of the rupture fault, the fault mechanism, the frequency-spectrum of wave energy released, the geological conditions, and the soils at a given site.

The most widely used scale historically was originated by Rossi and Forell in 1878. A later modification developed by Mercalli in Italy, now termed the **Modified Mercalli Intensity (MMI)** scale, is suitable for conditions in the United States. Bolt (2003) describes the details of the various intensity measures.

The geographical distribution of intensity is summarized by constructing isoseismal curves, or contour lines, which separate areas of equal intensity. The most probable position of the epicenter and the causative fault rupture is inside the area of highest intensity. An example of MMI curves for two moderate events is given in Figure 2-6. Clearly there can be large regional differences in MMI. Such variations in seismic wave attenuation are discussed in Section 2.6.1.

Correlations have been worked out between measured characteristics of the seismic waves and the reported Modified Mercalli intensity. A common one is that between the maximum ("peak") ground acceleration, A (centimeters per second squared), and the MM intensity, I. Such correlations are only broadly successful, particularly at the higher intensities. The description of the seismic waves for architectural and engineering purposes depends on a mixture of parameters, many of which are dependent on the frequency of the seismic waves. Nevertheless, because in many parts of the world instrumental measurements of ground

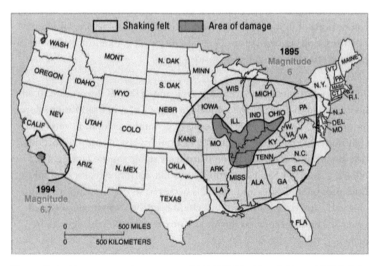

Figure 2-6: Map comparing curves of MMI 3 (shaking felt) and MMI 8 (area of damage) for the magnitude 6.7 1994 Northridge, California, earthquake and a magnitude 6 near New Madrid, Missouri, in 1895. Although the difference in magnitude implies an 11-fold difference in scalar seismic moment, the areas of shaking intensity for the smaller earthquake are substantially larger due to differences in seismic wave attenuation in the non-tectonic region of New Madrid compared to the western U.S. (discussed in section 2.6.1).

SOURCE: USGS FACT SHEET 017-03.

motion are not available, rough seismic intensity remains popular as a descriptor as well as for great historical earthquakes. Peak Ground Acceleration is employed as a measure in the current USGS **Shake-Maps** program, for example: these are maps showing ground shaking intensities that are available on the internet within a few minutes of an earthquake occurrence (see Section 2.6).

A number of other hazards of a geological nature may be triggered by an earthquake occurrence. These may at times cause severe damage and loss of life.

### ● Landslides

Landslides, ground settlement, and avalanches occur widely with and without earthquakes as a cause. All require special architectural treatment. Landslides and avalanches occur on slopes of a variety of geological materials. For engineering works, the speed at which a landslide develops and moves is a most important feature. Few defenses are available against rapid unexpected movements, but those that move slowly over periods of months to years lend themselves to some precautionary measures. Zoning regulations based on localized geological studies are the most effective mitigation measures.

During an earthquake, a series of seismic waves shakes the ground in all directions, so that under the critical conditions of water saturation, slope, and soil type, even relatively low levels of ground acceleration can cause a landslide. Even if these dynamic accelerations last for only a short time, widespread sliding can occur on marginally stable slopes. During and following the 1971 San Fernando, California, earthquake, for example, thousands of landslides and rockfalls occurred in the San Gabriel Mountains and caused a prominent dust-cloud over the strongly shaken area for days. This was repeated during the nearby 1994 Northridge earthquake.

Another human catastrophe caused by an earthquake-triggered debris avalanche occurred in Peru on May 31, 1970. The earthquake of magnitude 7.7 stimulated a rock avalanche amounting to some 50 million cubic meters of rock, snow, ice, and soil that travelled 15 km from the north peak of Huascarn Mountain, buried the towns around Ranraharca and most of Yungay, and killed at least 18,000 people.

In many instances, smaller landslides and avalanches can be detected in advance by suitable instrumentation installed on the slope with the readings monitored at regular intervals. Means of control can then be applied in appropriate circumstances: for example, removing small volumes of material to relieve the load at the head of the slope and adding material to the toe can be accomplished by earth-moving equipment. For cuts that are man-made, local regulations or ordinances may need to be developed and enforced during construction in a vulnerable area. Slopes made of fill, for example, may be required to be no steeper than 1 vertical to 1-1/2 horizontal, and the fraction of the soil covering the slope must be carefully controlled. Drainage of water away from such slopes is usually specified.

## ● Tsunamis and Seiches

The occurrence of an earthquake and a sudden offset along a major fault under the ocean floor, or a large submarine landslide, displaces the water like a giant paddle, thus producing powerful water waves at the ocean surface. When they reach a coastline, they may run up on land to many hundreds of meters. The elevation above the tide level (at the time of the tsunami) reached by the water is called the **run-up height**. This vertical distance is not the same as the tsunami water wave height offshore or the horizontal distance of water run-up from the normal water edge.

There have been tsunamis in most oceans of the world, but most notably in the Pacific Ocean. The coastline of Hilo, Hawaii, has seen inundation several times, and the giant earthquake in Alaska in 1964 had a run-up height of six meters in Crescent City, California, killing several people. Near the fault motion, 119 people drowned in Alaska.

A seismic sea wave warning system was set up in the Pacific after the devastating Aleutian tsunami of April 1, 1946. The tsunami warning center in Honolulu provides tsunami alerts and alerts local jurisdictions to issue warnings.

The best disaster prevention measures for a tsunami-prone coast involve zoning that controls the types and sizes of buildings that, if any, are permitted. If a site has a high possibility of tsunami incursion, the designer should consider some of the design provisions against flood, such as elevating the building above an estimated waterline. Of course in the case of locally generated tsunami, provisions must also be made for the severe strong shaking.

Long-period movements of water can also be produced in lakes and reservoirs by large earthquakes. These oscillations of lake levels are termed **seiches**. The November 2003 Denali earthquake in Alaska generated seismic seiches in wells and lakes of the south central United States. In the 1971 San Fernando, California, earthquake water sloshed out of swimming pools, producing some risk.

● Liquefaction

A notable hazard from moderate to large earthquakes is the liquefaction of water-saturated soil and sand produced by the ground shaking. In an earthquake, the fine-grained soil below the ground surface is subjected to alternations of shear and stress. In cases of low-permeability soils and sand, the water does not drain out during the vibration, building up pore pressure that reduces the strength of the soil.

Because earthquake shaking of significant amplitude can extend over large areas, and fine-grained soils in a saturated state are so widespread in their distribution, liquefaction has frequently been observed in earthquakes. In some cases, it is a major cause of damage and therefore is a factor in the assessment of seismic risk. Liquefaction in the 1964 Alaskan earthquake caused major disruptions of services and utilities and led to substantial building settlements and displacements. In the 1971 San Fernando, California, earthquake, liquefaction of soils in the San Fernando Dam caused a landslide in the upstream portion of the dam structure that almost resulted in a catastrophic dam failure. Widespread liquefaction resulted in severe damage after the 1811-1812 New Madrid and 1886 Charleston, South Carolina, earthquakes.

Many seismic regions have available liquefaction maps so that the risk of liquefaction at building sites can be assessed. Soil engineers have developed various technical methods of controlling liquefaction, the description of which goes beyond this chapter (see Chapter 3).

## 2.3 SEISMIC WAVES AND STRONG MOTION

### 2.3.1 Seismic Instrumental Recordings and Systems

Seismographs are instruments that are designed to record ground motions such as accelerations and displacements in earthquakes. Nowadays, technological developments in electronics have given rise to high-precision pendulum seismometers and sensors of both weak and strong ground motion. In these instruments, the electronic voltages produced by motions of a pendulum or the equivalent are passed through electronic circuitry to amplify the ground motion and digitize the signals for more exact measurements.

When seismic waves close to their source are to be recorded, special design criteria are needed. Instrument sensitivity must ensure that the relatively large amplitude waves remain on scale. For most seismological and engineering purposes, the wave frequency is high (1 to 10 Hz, i.e., cycles per second), so the pendulum or its equivalent can be small. For comparison, displacement meters need a pendulum with a long free period (many seconds).

Because many strong-motion instruments need to be placed at unattended sites for periods of months or years before a strong earthquake occurs, they usually record only when a trigger mechanism is actuated with the onset of seismic motion. Solid-state memories are now used with digital recording instruments, making it possible to preserve the first few seconds before the trigger starts the permanent recording. In the past, recordings were usually made on film strips, providing duration of up to a few minutes.

In present-day equipment, digitized signals are stored directly on a memory chip, and are often telemetered to central recording sites in near real-time (several to tens of seconds). In the past, absolute timing was not provided on strong-motion records but only accurate relative time marks; the present trend, however, is to provide Universal (Greenwich Mean) Time - the local mean time of the prime meridian by means of special radio receivers or Global Positioning Satellite (GPS) receivers.

The prediction of strong ground motion and response of engineered structures in earthquakes depends critically on measurements of the locational variation of earthquake intensities near the fault. In an effort to secure such measurements, special arrays of strong-motion seismographs have been installed in areas of high seismicity around the world, both away from structures **(free field)** and on them (Figure 2-7). The seismic instrumentation of various types of buildings is clearly to be encouraged by architects, both for post-earthquake performance evaluation, future design modification and improved emergency response.

It is helpful for the user of strong-motion **seismograms** (called "**time histories**") to realize that the familiar "wiggly line" graphic records are not the actual motion of the ground, but have been filtered in some way by both the recording instrument and by the agency providing the data (see Section 2.6). In most cases, however, for practical applications the architect or engineer need not be concerned about the difference.

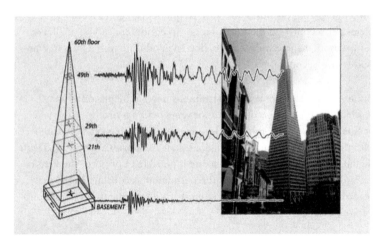

Figure 2-7: Transamerica "Pyramid" building in downtown San Francisco.

Modern instruments capable of recording large motions strategically placed in structures provide information on the structural response. In this case it is evident that there is amplification of both short-period and long-period motions in the upper floors. Also the duration of shaking at periods corresponding to characteristic vibrations of the structure become quite long towards the top.

SOURCE: USGS FACT SHEET 017-03.

NATURE OF EARTHQUAKES AND SEISMIC HAZARDS

## 2.3.2 Types of Earthquake Waves

In most instances of seismic ground motions in solid rock or soil, the waves involved are made up of four basic types of elastic waves that create the shaking that people feel and that causes damage in an earthquake. These waves are similar in many important ways to the waves observed in air, water, and elastic solids.

The first two types of waves travel through the body of the earth before arriving at the surface. The faster of these "body" waves is appropriately called the **primary** or P wave (Figure 2-8a). Its motion is the same as that of a sound wave in that, as it spreads out, it alternately pushes (compresses) and pulls (dilates) the rock. These **P** waves, just like acoustic waves, are able to travel through solid rock, such as granite and alluvium, through soils, and through liquids, such as volcanic magma or the water of lakes and oceans.

The second and slower seismic body wave through the earth is called the **secondary** or **S** wave or sometimes the **shear wave** (Figure 2-8b). As an S wave propagates, it shears the rocks sideways at right angles to the direction of travel. At the ground surface, the upward emerging S waves also produce both vertical and horizontal motions. Because they depend on elastic shear resistance, S waves cannot propagate in liquid parts of the earth, such as lakes. As expected from this property, their size is significantly weakened in partially liquefied soil. The speed of both P and S seismic waves depends on the density and elastic properties of the rocks and soil through which they pass. In earthquakes, P waves move faster than S waves and are felt first. The effect is similar to a sonic boom that bumps and rattles windows. Some seconds later, S waves arrive with their significant component of side-to-side shearing motion. As can be deduced from Figure 2-8, for upward wave incidence, the ground shaking in the S waves becomes both vertical and horizontal, which is the reason that the S wave motion is so effective in damaging structures.

The other two types of earthquake waves are called **surface waves** because their motion is restricted to near the earth's surface. Such waves are analogous to waves in the ocean that do not disturb the water at depth. In a similar way, as the depth below the ground surface increases, the ground displacements of seismic surface waves decrease.

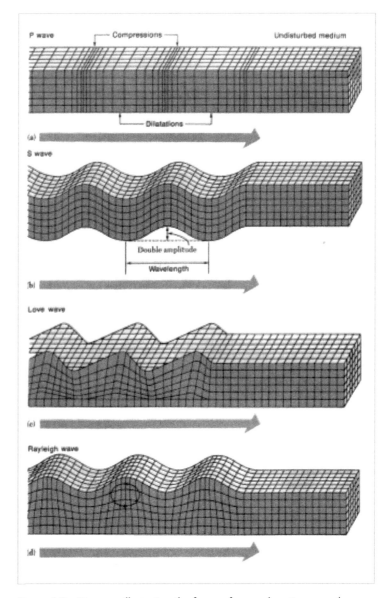

Figure 2-8: Diagram illustrating the forms of ground motion near the ground surface in four types of earthquake waves.

The first type of surface wave is called a **Love wave** (Figure 2-8c) Its motion is the same as that of S waves that have no vertical displacement; it moves the ground side to side in a horizontal plane parallel to the earth's surface, but at right angles to the direction of propagation. The second type of surface wave is called a **Rayleigh wave** (Figure 2-8d). Like ocean waves, the particles of rock displaced by a Rayleigh wave move both vertically and horizontally in a vertical plane oriented in the direction in which the waves are traveling. The motions are usually in a retrograde sense, as shown by the arrows in Figure 2-8. Each point in the rock moves in an ellipse as the wave passes.

Surface waves travel more slowly than P and S waves and Love waves travel faster than Rayleigh waves in the same geological formation. It follows that as the seismic waves radiate outwards from the rupturing fault into the surrounding rocks, the different types of waves separate out from one another in a predictable pattern. However, because large earthquake fault sources have significantly extended slip surfaces (i.e., many tens of kilometers), the separation is often obscured by overlapping waves of different wave types at sites close to the fault. Examples of near-fault large amplitude time histories are shown in Figure 2-9.

As seismic body waves (the P and S waves), move through layers of rock or soil, they are reflected or refracted at the layer interfaces. To complicate matters further, whenever either one is reflected or refracted, some of the energy of one type is converted to waves of the other type. When the material stiffnesses differ from one layer to another, the layers act as wave filters that amplify the waves at some frequencies and deamplify them at others.

It is important to note that when P and S waves reach the surface of the ground, most of their energy is reflected back into the crust, so that the surface is affected almost simultaneously by upward and downward moving waves. For this reason, considerable amplification of shaking typically occurs near the surface, sometimes doubling the amplitude of the upcoming waves. This surface amplification enhances the input shaking to structures and is responsible for much of the damage produced at the surface of the earth. In contrast, in many earthquakes, mineworkers below ground report less shaking than people on the surface. Nowadays, it is routine for soil engineers to make allowance for the wave amplification effect as the input seismic waves pass upwards through the soil layer to the ground surface.

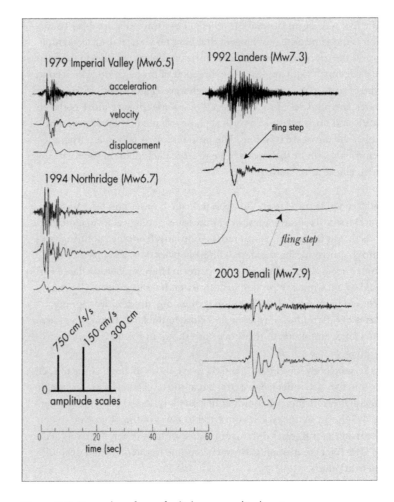

Figure 2-9: Examples of near-fault, large amplitude seismograms (time-histories).

The figure includes records from Imperial Valley, Landers (Lucerne), Northridge (Newhall) and Denali (Trans-Alaska Pipeline). Note the permanent offset in displacement of the Landers record. This is due to fault ground rebound or fling, shown by the arrows. The bars (lower left) give the common amplitude scales for the displacement, velocity and acceleration records.

NATURE OF EARTHQUAKES AND SEISMIC HAZARDS

It should be noted that seismic S waves travel through the rocks and soils of the earth with both a shearing and a rotational component. The latter components of ground motion have important effects on the response of certain types of structures, and some building codes now take rotational ground motion into consideration.

Seismic waves of all types progressively decrease in amplitude with distance from the source. This attenuation of waves varies with different regions in the United States. The attenuation of S waves is greater than that of P waves, but for both types attenuation increases as wave frequency increases. Ground motion attenuation can flatten and even reverse its downward trend due to strong reflected arrivals from rock interfaces. It has been shown that such reflections led to elevated ground motions in the 60-80 km distance range from the 1989 Loma Prieta, California, earthquake (i.e., in Oakland and San Francisco). Deposits of low velocity sediments in geological basins can also cause elevated levels of ground motions.

For a more detailed discussion of seismic wave attenuation and theoretical wave amplitude, see Section 2.6.1.

The physical characteristics of seismic waves have been verified by many recordings at moderate (15-30 km) to larger distances from the wave source called the **far-field**, but are not adequate to explain important details of the heavy shaking near the source of an energetic earthquake called the **near-field**. As explained above, near a rupturing fault, the strong ground shaking consists of mixtures of seismic wave types that have not separated distinctly. Although this complication makes identification of P, S, and surface waves on strong motion records obtained near the rupturing fault difficult, there has been recent progress in this skill, based on correlations between actual recordings and theoretical modelling. This advance has made possible the computation of realistic ground motions at specified sites for engineering design purposes.

Three final points about seismic waves are worth emphasizing here:

○ Earthquake waves are much affected by soil elastic properties. For example, in weathered surface rocks, alluvium and water-saturated soil, the relative sizes of P, S, and surface waves can vary significantly, depending on wave frequency, as they propagate through the surficial non-homogenous geological structures. Under extreme conditions of large wave amplitude and special geotechnical properties, the linear elastic behavior breaks down and nonlinear effects occur.

○ Patterns of incoming seismic waves are modified by the three-dimensional nature of the underground geological structures. As mentioned above, instrumental evidence on this effect was obtained from recordings of the 1989 Loma Prieta, California, earthquake. In this case, strong-motion recordings indicated that there were reflections of high-frequency S-waves from the base of the earth's crust at a depth of about 25 km under the southern San Francisco Bay. Also, in this earthquake, large differences in the rock structure from one side of the San Andreas fault to the other produced variations in ground motion by lateral refraction of S waves. The effect produced significant S wave amplitude variation as a function of azimuth from the seismic source, in a period range of about 1 to 2 seconds. In addition, there was measurable scattering of S waves by separate alluvial basins in the south part of San Francisco Bay. Overall, the seismic intensity was enhanced in a region between San Francisco and Oakland, about 10 km wide by 15 km long. The observed damage and seismic intensity are well explained by these seismological results.

○ It is important to explain the special seismic intensity enhancement in the near field of the earthquake source. Because of special features of engineering importance, this discussion of seismic wave patterns near to the fault source is given in the separate Section 2.4. As may be seen in Figure 2-10, time histories of the seismic waves contain pulse-like patterns of motion of crucial importance to earthquake response of larger structures.

## 2.4. SEISMIC SOURCES AND STRONG MOTION

As has been discussed in the previous sections, seismic waves are generally generated by the sudden rupture of faults, but can also be initiated by other natural processes, such as pulsing of volcanic magma and landsliding. They can also be caused by man-made explosions and collapse of subterranean mines. The strength of S-wave radiation depends upon the mechanism of the source. In particular, fault rupture is an efficient generator of S waves, which are responsible for much of the demand of earthquakes on the built environment. The seismic wave amplitudes vary

NATURE OF EARTHQUAKES AND SEISMIC HAZARDS

with azimuth from the source as a result of the orientation of the force couples that cause the fault rupture. The resulting pattern of radiation of all types of seismic waves may be described mathematically using the same terms used in defining the different types of faults (see Figure 2-3), i.e., in terms of the fault strike, dip, and direction of slip.

## 2.4.1 Earthquake Magnitude

The original instrumental measure of earthquake size has been significantly extended and improved in recent years. First, because the fundamental period of the now superseded Wood-Anderson seismograph is about 0.8 sec., it selectively amplifies those seismic waves with periods ranging from 0.5 to 1.5 sec. It follows that because the natural periods of many building structures are within this range, the first commonly used parameter, called the **Richter magnitude** ($M_L$) based on this seismograph, remains of value to architects. Generally, shallow earthquakes have to attain Richter magnitudes of more than 5.5 before significant damage occurs, even near the source of the waves. It should be remembered that a one unit increase in magnitude indicates a tenfold increase in the amplitude of the earthquake waves.

The definition of all magnitude scales entails that they have no theoretical upper or lower limits. However, the size (i.e., the seismic moment) of an earthquake is practically limited at the upper end by the strength of the rocks of the earth's crust and by the area of the crucially strained fault source. Since 1935, only a few earthquakes have been recorded on seismographs that have had a magnitude over 8.0 (see Table 2-1). At the lower extreme, highly sensitive seismographs can record earthquakes with a magnitude of less than minus two.

For reference, an architect may still encounter the following magnitude scales.

○ **Surface Wave Magnitude** ($M_s$) is based on measuring the amplitude of surface waves with a period of 20 sec. Surface waves with a period around 20 sec are often dominant on the seismograph records of distant earthquakes (epicentral distances of more than 1,000 km).

○ **Body Wave Magnitude** ($M_b$) Because deep focus earthquakes have no trains of surface waves, only the amplitude of the recorded P wave is used.

Nowadays, because of the shortcomings of $M_L$, $M_b$, and to a lesser degree $M_s$ in distinguishing between the size of the biggest earthquakes, the **Moment Magnitude** scale, $M_w$, has replaced earlier definitions.

Studies have shown that the Richter Magnitude ($M_L$) scale progressively underestimates the strength of earthquakes produced by large fault ruptures. The upper-bound value for this scale is about $M_L = 7$. The body wave magnitude ($M_b$) saturates at about the same point. In contrast, the surface-wave magnitude ($M_s$) that uses the amplitude of waves with periods of 20 seconds saturates at about $M_s = 8$. Its inadequacy in measuring the size of great earthquakes can be illustrated by comparing values for the San Francisco earthquake of 1906 and the great Chilean

Figure 2-10: Two earthquakes may have equal magnitudes but be distinctly unequal in other respects.

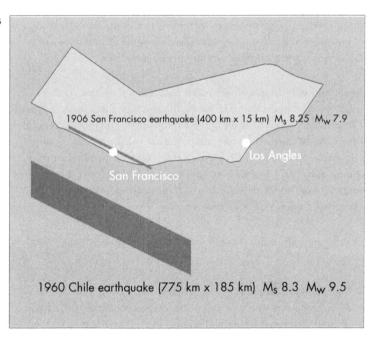

1906 San Francisco earthquake (400 km x 15 km)  $M_s$ 8.25  $M_w$ 7.9

Los Angles

San Francisco

1960 Chile earthquake (775 km x 185 km)  $M_s$ 8.3  $M_w$ 9.5

The 1906 San Francisco, California, earthquake ruptured rock over a shorter length and shallower depth - only about 1/25 the area - as the 1960 Chilean earthquake. Although the surface wave magnitudes are the same, the moment magnitude for these two earthquakes (Table 2-1) are distinctly different. A sketch of the outline of California is shown for scale.

NATURE OF EARTHQUAKES AND SEISMIC HAZARDS

earthquake of 1960. Both earthquakes had a surface wave magnitude ($M_s$) of 8.3. However, the area that ruptured in the San Francisco earthquake was approximately 15 km deep and 400 km long, whereas the length that ruptured in the Chilean earthquake was equal to about half of the state of California. Clearly the Chilean earthquake was a much "larger" event (Figure 2-10).

The moment-magnitude scale ($M_W$) does not suffer from saturation for great earthquakes. The reason is that it is directly based on the forces that work over the area of the fault rupture to produce the earthquake and not on the amplitude and limited frequencies of specific types of seismic waves. Hence, as can be expected, when moment magnitudes were assigned to the 1906 San Francisco earthquake and the 1960 Chilean earthquake, the magnitude of the San Francisco earthquake dropped to 7.9, whereas the magnitude of the Chilean earthquake rose to 9.5. $M_S$ and $M_W$ for some great earthquakes are compared in Table 2-1.

## 2.4.2 Elastic Rebound and its Relationship to Earthquake Strong Ground Motion

The slip along the San Andreas fault that produced the 1906 earthquake was studied by H. F. Reid. He imagined a bird's-eye view of a straight line drawn at a certain time at right angles across the San Andreas fault. As the tectonic force slowly works, the line bends, the left side shifting in relation to the right. The deformation amounts to about a meter in the course of 50 years or so. This straining cannot continue indefinitely; sooner or later the weakest rocks, or those at the point of greatest strain, break. This fracture is followed by a springing back or **rebounding**, on each side of the fracture.

This **elastic rebound** was believed by Reid to be the immediate cause of earthquakes, and his explanation has been confirmed over the years. Like a watch spring that is wound tighter and tighter, the more the crustal rocks are elastically strained, the more energy they store. When a fault ruptures, the elastic energy stored in the rocks is released, partly as heat and partly as elastic waves. These waves are the earthquake. A remarkable example of this phenomenon that produced striking offsets occurred in Turkey in the 1999 Izmit earthquake (Figure 2-4).

Straining of rocks in the vertical dimension is also common. The elastic rebound occurs along dipping fault surfaces, causing vertical disruption

Table 2-1: Magnitudes of some great earthquakes

| Date | Region | $M_S$ | $M_W$ |
|---|---|---|---|
| January 9, 1905 | Mongolia | 8.25 | 8.4 |
| January 31, 1906 | Ecuador | 8.6 | 8.8 |
| April 18, 1906 | San Francisco | 8.25 | 7.9 |
| January 3, 1911 | Turkestan | 8.4 | 7.7 |
| December 16, 1920 | Kansu, China | 8.5 | 7.8 |
| September 1, 1923 | Kanto, Japan | 8.2 | 7.9 |
| March 2, 1933 | Sanrika | 8.5 | 8.4 |
| May 24, 1940 | Peru | 8.0 | 8.2 |
| April 6, 1943 | Chile | 7.9 | 8.2 |
| August 15, 1950 | Assam | 8.6 | 8.6 |
| November 4, 1952 | Kamchatka | 8 | 9.0 |
| March 9, 1957 | Aleutian Islands | 8 | 9.1 |
| November 6, 1958 | Kurile Islands | 8.7 | 8.3 |
| May 22, 1960 | Chile | 8.3 | 9.5 |
| March 28, 1964 | Alaska | 8.4 | 9.2 |
| October 17, 1966 | Peru | 7.5 | 8.1 |
| August 11, 1969 | Kurile Islands | 7.8 | 8.2 |
| October 3, 1974 | Peru | 7.6 | 8.1 |
| July 27, 1976 | China | 8.0 | 7.5 |
| August 16, 1976 | Mindanao | 8.2 | 8.1 |
| March 3, 1985 | Chile | 7.8 | 7.5 |
| September 19, 1985 | Mexico | 8.1 | 8.0 |
| September 21, 1999 | Taiwan | 7.7 | 7.6 |
| November 2, 2002 | Alaska | 7.0 | 7.9 |
| December 26, 2004 | Sumatra | NA | 9.0 |

NATURE OF EARTHQUAKES AND SEISMIC HAZARDS

in level lines at the surface and fault scarps. Vertical ground displacement too can amount to meters in dip-slip faulting (as in the 1999 Chi Chi, Taiwan, earthquake, faulting in Figure 2-5).

Observations show that fault displacement occurs over a continuum of rates from less than a second to very slow fault slip. Although the latter "creep" can pose significant hazard for structures built across such rupturing faults, these slow slips do not radiate elastic seismic waves. Indeed, the generation of strong seismic waves requires that the elastic rebound of the fault is rapid. The Lucerne record (Figure 2-9) for the Landers earthquake 3 km from the fault shows that elastic rebound (**fling-step**) occurred over about 7 seconds. This static offset arises from near-field waves and their amplitudes attenuate more rapidly than far-field body waves. Since this attenuation is strong, and the rise time of the fling-step increases with distance, large dynamic motions derived from this phenomenon are typically limited to sites very close to the fault. A time derivative of the fling-step produces a pulse-like velocity record (Figure 2-10). For example the fling-step at the Lucerne site for the Landers earthquake was recorded 3 km from the fault trace, and the 3 m/s peak velocity recorded within 1 km of the fault for the 1999 Chi-Chi, Taiwan, earthquake (Table 2-2) had a significant contribution from the fling-step.

## 2.4.3 Source Directivity and its Effect on Strong Ground Motions

For structures near active faults an additional seismological source effect may be important in design in which the direction and speed of a rupture along a fault focuses wave energy, producing direction-dependent seismic wave amplitudes. This direction-dependent amplitude variation called **directivity** affects the intensity and damage potential of strong ground motions near and at moderate distances from the fault source. In contrast to large pulse-like dynamic motions derived from a fling-step, those due to directivity are results of the superposition or focusing of far-field body waves. Since waves distant from the fault attenuate less with distance than those nearby, directivity pulses with elevated motions can occur some distance from the fault. To keep these two effects separate, the terms "**directivity pulse**" and "**fling-step**" have been used for the rupture directivity and elastic rebound effects, respectively.

Directivity is a term that describes the focusing and defocusing of waves due to the direction of rupture with respect to the direction to a given site. Therefore it describes azimuthal variation in earthquake ground motion about the fault. The difference between the rupture direction and the direction to the site is related by an angle. Large ground accelerations and velocities are associated with small angles, since a significant portion of the seismic energy is channeled in the direction to the site. Consequently, when a large urban area is located within the small angle, it will experience severe damage. Studies show that in the Northridge earthquake of 1994, the rupture propagated in the direction opposite from downtown Los Angeles and the San Fernando Valley, causing only moderate damage, whereas the collapsed SR-18/I5 highway interchange was in an area of small angle. In the Kobe, Japan, earthquake of 1995, the rupture was directed towards the city of Kobe, resulting in severe damage. The stations that lie in the direction of the earthquake rupture propagation will record shorter strong-motion duration than those located opposite to the direction of propagation.

Directivity can significantly affect strong ground motion by as much as a factor of 10, and methods are being developed to account for this effect through numerical simulation of earthquake ground motions, and by empirical adjustment of ground motion attenuation relationships. However, it is not clear how to incorporate directivity into methods for predicting ground motion in future earthquakes, because the angle between the direction of rupture propagation and the source to recording site and the slip history on the fault is not known before the earthquake. Studies that incorporate directivity in the analysis must therefore investigate many rupture scenarios to examine the range of possible motions.

## 2.5 STRONG GROUND MOTION

As mentioned earlier, for architectural purposes it is important to know that near-fault ground motions often contain significant **velocity wave pulses**, which may be from fling in the near-fault, fault-parallel direction, or from directivity in the fault-normal direction extending a considerable distance from the ruptured fault. For strike-slip fault sources, they dominate the horizontal motion and may appear as single or double pulses, each with single or double-sided amplitudes. The duration (period) of the main pulse may range from 0.5 sec. to 5 sec. or more for the greatest magnitudes. These properties depend on the type, length, and complexity of the fault rupture.

## 2.5.1 Duration of Strong Shaking

Field studies demonstrate that the duration of strong ground shaking is often a critical factor in the response of foundation materials and structures. There is no way to determine the duration of a design event and factor duration into current design codes. Soil response in particular can be strongly dependent on the increases in pore water pressure with repeated cyclic input. Also nonlinear degradation of damaged structures (also caused by long shaking and in large aftershocks) can lead to collapse.

## 2.5.2 Estimating Time Histories

Numerical modeling can be particularly helpful in predicting the effect of certain special geological structures on a hazard at a site. Consider, for example, the response of the Los Angeles alluvial basin to a large earthquake from slip of the San Andreas fault. A computer simulation was made in 1995 by Olsen et al. that gives wave motion for a three-dimensional numerical model, when the source is a magnitude 7.75 earthquake along the 170 km section of the San Andreas fault between Fort Tejon Pass and San Bernardino. The results are graphed in Figure 2-11. The wave propagation is represented as horizontal velocities of the ground parallel to the San Andreas fault.

The snapshots show that after 40 sec., ground motion in the basin begins to intensify, and 10 sec later the entire Los Angeles basin is responding to large amplitude surface waves. (The waves shown are spectrally limited to frequencies below 0.4 Hz. In an actual earthquake, the ground motions would contain much higher frequencies, but the effects would be similar.) The component of motion perpendicular to the fault strike is 25% larger than the parallel component near the fault due to the directivity of the rupture (see Section 2.4.2). This simulation predicted long-period peak ground velocities greater than 1 m/sec. at some areas in Los Angeles, even though the main trough of the basin is about 60 km from the fault. Later analysis of the same region suggests that such computed amplitude factors (up to six in deeper parts of the basin) should be used by planners and designers as a guide only and with caution.

Instead of such synthetic models, quasi-empirical seismic strong ground motions, based on modified actual recordings of similar earthquakes, are now normally used to estimate seismic hazard. Two equivalent representations of the hazard are commonly considered together. The first is an

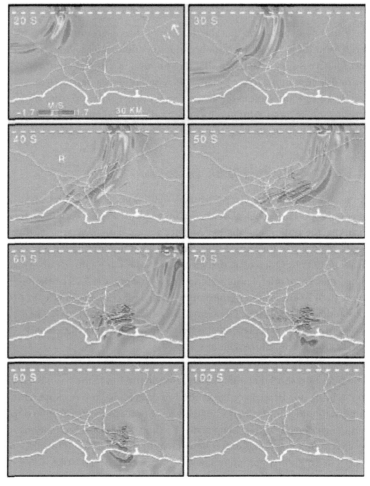

Figure 2-11: Aerial snapshots of a simulated wave propagation in the Los Angeles area.

The snapshots depict velocities from 20 s to 100 s after the origin time of the rupture. Red depicts large amplitudes of both positive and negative polarity. R depicts the initiation of an area of local resonance above the deepest part of the San Fernando basin. The particle motion is scaled by a constant for all snapshots.

SOURCE: OLSEN ET AL. (1995) FOR A HYPOTHETICAL SAN ANDREAS FAULT EARTHQUAKE

NATURE OF EARTHQUAKES AND SEISMIC HAZARDS

estimate of the **time-history** of the ground motion appropriate to the site. The second is the **response spectra** (the spectral response of a damped single degree-of-freedom harmonic oscillator, see section 4.5.2) for the whole seismic motion at the site. These two representations of seismic hazard can be connected by appropriate transformations between the time and frequency descriptions of the earthquake.

In the simplest time-history representation, the major interest of architects and engineers in assessing the earthquake risk has traditionally been in the peak ground acceleration (PGA), velocity, and displacement as a function of frequency, or period. In recent work related to large and critical engineered structures, however, the pattern of wave motion has been recognized as crucial in structural response, because the nonlinear response of such structures is often dependent on the sequence of arrival of the various types of waves. In other words, damage would be different if the ground motion were run backwards rather than in the actual time sequence of arrival. The sequence (phasing) of the various wave types on the artificial seismograms can be checked from seismological knowledge of times of arrival of the P, S, directivity-pulse, fling, and surface waves. Only in this way can a realistic envelope of amplitudes in the time histories be assumed.

In the usual calculation, the initial step is to define, from geological and seismological information, the fault sources that are appropriate and dangerous for the site of interest. This fault source selection may be largely deterministic, based on prior experience, or largely probabilistic, and may be decided on grounds of acceptable risk. Next, specification of the propagation path of the strongest waves is made, as well as the P, S and surface wave velocities along the path. These speeds allow calculation of the appropriate delays in wave propagation between the source and the multi-support points of the structure and the angles of approach of the incident seismic waves.

The computation of realistic motions then proceeds as a series of nonlinear iterations, starting with the most appropriate observed strong-motion record available, called the seed motion, to a set of more specific time histories, which incorporate the seismologically defined wave patterns. The seed strong-motion accelerograms are chosen to approximate the seismic source type (dip-slip, etc.) and geological specifications for the region in question. (A set of suggested time histories for seed motions is listed in Table 2-2). Many sample digitized records can be downloaded using the Virtual Data Center (VDC) website of the Consortium of Organizations for Strong-Motion Observational Systems (COSMOS) (see section 2.10). The frequency content of the desired time-history is controlled by applying engineering constraints, such as a selected response amplitude spectrum. Such target spectra are obtained, for example, from previous engineering analysis and from earthquake building codes (see, e.g., IBC, 2003).

## 2.6. SEISMIC HAZARD

### 2.6.1 Empirical Attenuation Curves

As has been outlined in the previous sections, the estimation of the earthquake hazard in a region or at a site requires the prediction of ground motions. The empirical estimation of seismic hazard curves is a necessary step. It follows that hazard calculations involve a number of assumptions and extrapolations. The common initial difficulty is ignorance of the actual seismic wave attenuation for the site in question, despite the recent publication of a variety of average curves for certain regions. The importance of attenuation factors in calculation of predicted ground motion at arbitrary distances has led to competing empirical attenuation forms based on available intensity measurements and geological knowledge.

Usually wave attenuation changes significantly from one geological province to another, and local regional studies are advisable to calibrate the parameters involved.

As mentioned in Section 2.4, although different measures of earthquake magnitude are still used, particularly with historical data, the moment magnitude ($M_w$) is now usually adopted as a standard measure of size in attenuation statistics. Also, nowadays, some form of "closest" distance to the rupture is used as the distance parameter rather than epicentral or hypocentral distance. It is important to use the appropriate distance measure for a given attenuation relation. The most common source, ray path, and site parameters are magnitude, distance, style-of-fault, directivity, and site classification. Rupture directivity is defined in detail in Section 2.4.3 and is not discussed here. In some studies, additional parameters are used: hanging-wall flag, rupture directivity parameters, focal depth, and soil depth classification.

Table 2-2: Examples of near-fault strong-motion recordings from crustal earthquakes with large peak horizontal ground motions

| Earthquake | Magnitude $M_W$ | Source Mechanism | Distance km* | Acceleration (g) | Velocity (cm/sec) | Displace (cm) |
|---|---|---|---|---|---|---|
| 1940 Imperial Valley    (El Centro, 270) | 7.0 | Strike-Slip | 8 | 0.22 | 30 | 24 |
| 1971 San Fernando    (Pacoima 164) | 6.7 | Thrust | 3 | 1.23 | 113 | 36 |
| 1979 Imperial Valley    (EC #8, 140) | 6.5 | Strike-Slip | 8 | 0.60 | 54 | 32 |
| Erzican    (Erzican, 1992) | 6.9 | Strike-Slip | 2 | 0.52 | 84 | 27 |
| 1989 Loma Prieta    (Los Gatos, 000) | 6.9 | Oblique | 5 | 0.56 | 95 | 41 |
| 1992 Lander    (Lucerne, 260) | 7.3 | Strike-Slip | 1 | 0.73 | 147 | 63 |
| 1992 Cape Mendocino    (Cape Mendocino, 000) | 7.1 | Thrust | 9 | 1.50 | 127 | 4 |
| 1994 Northridge    (Rinaldi, 228) | 6.7 | Thrust | 3 | 0.84 | 166 | 29 |
| 1995 Kobe    (Takatori, 000) | 6.9 | Strike-Slip | 1 | 0.61 | 127 | 36 |
| 1999 Kocaeli    (SKR, 090) | 7.4 | Strike-Slip | 3 | 0.41 | 80 | 205 |
| 1999 Chi-Chi    (TCU068, 000) | 7.6 | Thrust | 1 | 0.38 | 306 | 940 |

* distence km shows surface distance from fault

There are also differences in site classification schemes in different regions that make comparison of estimates of ground motions from alternative estimates difficult. Broad site categories such as "rock," "stiff-soil," and "soft-soil" are common and affect ground motions (Figure 2-12), but more quantitative site classifications based on the S-wave velocity, such as the average S-wave speed in the top 30 m, are now preferred. Most attenuation relations simply use a site category such as "deep soil"; however, this general category covers a wide range of soil depths from less than 100 m to several kilometers of sediments. Some attenuation relations use an additional site parameter to describe the depth of the sediment.

For thrust faults, high-frequency ground motions on the upthrown block (hanging-wall side of a thrust fault) are much larger than on the downdropped block (footwall). This increase in ground motions on the hanging wall side is in part an artifact of using a rupture distance measure, but may also be due to the dynamics of waves interacting with the dipping fault plane and the surface of the earth. If a site on the hanging wall and footwall are at the same rupture distance, the site on the hanging wall side is closer to more of the fault than the site on the footwall side. Such difference was marked in damage patterns to houses and other structures in the 1999 Chi Chi, Taiwan, earthquake ($M_W = 7.6$).

In the eastern U.S., incorporation of a variation in the distance slope of the attenuation relation to accommodate the increase in ground motions due to supercritical reflections from the base of the crust has been suggested. Typically, this result leads to a flattening of the attenuation curve at distances of about 100 km). This is most significant for regions in which the high activity sources are at a large distance from the site.

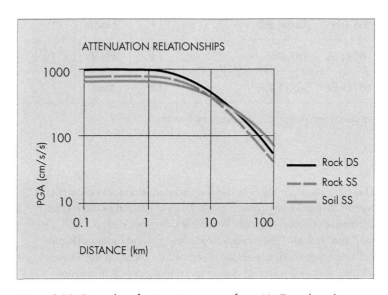

Figure 2-12: Examples of attenuation curves for a $M_W7$ earthquake obtained by data regression, illustrating the effects of a **site** type: rock (blue dashed) vs. deep soil (red), and **event** type: strike-slip fault (blue dashed) vs. reverse fault (black)

SOURCE: ABRAHAMSON AND SILVA, 1997

An important statistical issue in developing attenuation relations is the uneven sampling of the data from different earthquakes. For example, in some cases, an earthquake may have only one or two recordings (e.g., the 1940 El Centro event), whereas, some of the recent earthquakes have hundreds of recordings (e.g., the 1999 Chi Chi earthquake). The use of statistical weights can reduce this uneven sampling problem. There are two extremes: give equal weight to each data point or give equal weight to each earthquake. The random-effects model seems best. It uses a weighting scheme that varies between giving equal weight to each earthquake and equal weight to each data point, depending on the distribution of the data.

In addition to the median measure of ground motion, the standard deviation of the measured ground motion parameters is also important for either deterministic or probabilistic hazard analyses. Worldwide, it is common to use a constant standard deviation, but recently, several attenuation relations have attributed magnitude or amplitude dependence to the standard deviation.

## 2.6.2 Probabilistic Seismic Hazard Analysis (PSHA) and Building Codes

Probabilistic Seismic Hazard Analysis provides an estimate of the likelihood of hazard from earthquakes based on geological and seismological studies. It is probabilistic in the sense that the analysis takes into consideration the uncertainties in the size and location of earthquakes and the resulting ground motions that could affect a particular site. Seismic hazard is sometimes described as the probability of occurrence of some particular earthquake characteristic (such as peak ground acceleration) For statistical reasons, these probabilities cover a range of values, and because risk involves values being greater than expected, the word "exceedance" has been coined as explained below.

Probabilistic analysis uses four basic steps in order to characterize the probable seismic hazard:

● Identification of the seismic source or faults.

This often includes the identification of surface faulting features that can be recognized as active. Seismic sources may be specified as site specific, for an active source region or, when geologic information is poor, for random occurrence of active faults in the study region. Once the faulting hazard is identified, earthquake occurrence statistics are compiled, which might be in the form of annual rates of seismic events or, in an active regions of known faults, more specific information provided by paleoseismic studies such as dating episodes of fault offsets. (Paleoseismology involves digging to expose the underground face of a fault, so that historic offsets can be made visible and material suitable for radiocarbon age dating can be obtained). The objective is to obtain a measure of the frequency of earthquakes within a given time period as a function of magnitude that may be expressed as a probabilistic statement (or mathematical likelihood) of the earthquake occurrence.

● Characterization of annual rates of seismic events.

As an example, if there is one magnitude 7 earthquake in a given region every 50 years, then the annual rate of occurrence is 0.5. Commonly used maps to express probability are cast in terms of a 50-year return period, and are used to determine the ground motion values to be specified in building codes and used in seismic design.

Since damaging ground motions can result from nearby moderate earthquakes as well as large distant earthquakes, the recurrence rates for each magnitude range must be determined.

● Development of attenuation relationships

Attenuation relationships and their uncertainty due to limited information must be developed so that the ground motion parameters for each of the sources developed in the first step can be related to the distance of the study site from them.

● Combining factors

The annual recurrence and the attenuation are combined to determine the site-specific hazard.

Until the 1990s, seismic building codes used a single map of the United States that divided the country into numbered seismic zones (0,1,2,3,4) in which each zone was assigned a single acceleration value in % g which was used to determine seismic loads on the structure.

Starting in the 1970s, new hazard maps began to be developed on a probabilistic basis. In the 1994 *NEHRP Recommended Provisions* (FEMA 222A), two maps of the US were provided in an appendix for comment. They showed effective peak acceleration coefficients and effective peak velocity–related coefficients by use of contour lines that designated regions of equal value. The ground motions were based on estimated probabilities of 10% of exceedance in various exposure times (50, 100 and 250 years). The 1997 *Recommended Provisions* (FEMA 302) provided the first spectral response maps to pass consensus ballot. This lead to the current maps which, with some revisions, are now used in the 2003 *NEHRP Provisions* (FEMA 450), the ASCE *Prestandard and Commentary for the Seismic Rehabilitation of Buildings* (FEMA 356), and the *International Building Code.*

The probabilistic analysis is typically represented in maps in the form of a percentage probability of exceedance in a specified number of years. For example, commonly used probabilities are a 10% probability of exceedance in 50 years (a return period of about 475 years) and a 2% probability of exceedance in 50 years (a return period of about 2,500 years). These maps show ground motions that may be equaled but are not expected to be exceeded in the next 50 years: the odds that they will not be exceeded are 90% and 98%, respectively.

Seismic hazard probability maps are produced by the United States Geological Survey (USGS) as part of the National Seismic Hazard Mapping Project in Golden, Colorado. The latest sets of USGS of maps provide a variety of maps for Peak Ground Acceleration and Spectral Acceleration, with explanatory material, and are available on the USGS web site

The USGS map shown in Figure 2-13 is a probabilistic representation of hazard for the coterminous United States. This shows the spectral acceleration in %g with a 2% probability of exceedance in 50 years: this degree of probability is the basis of the maps used in the building codes.

The return period of 1 in 2,500 years may seem very infrequent, but this is a statistical value, not a prediction, so some earthquakes will occur much sooner and some much later. The design dilemma is that if a more frequent earthquake - for example, the return period of 475 years - were used in the lower seismic regions, the difference between the high and low-probability earthquakes is a ratio of between 2 and 5. Design for

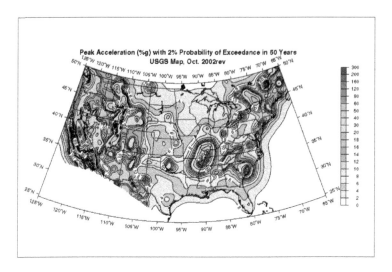

Figure 2-13: Spectral Acceleration values in %g with a 2% Probability of Exceedance in 50 5ears for the coterminous United States. The color scale to the right relates to the %g values.

SOURCE: USGS NATIONAL SEISMIC HAZARD MAPPING PROJECT

the high-probability earthquake would be largely ineffective when the low-probability event occurred

In practical terms, the building designer must assume that the large earthquake may occur at any time. Thus, use of the 2,500 return period earthquake in the lower seismic regions ensures protection against rare earthquakes, such as the recurrence of the 1811-1812 earthquake sequence in New Madrid, Missouri, or the 1898 Charleston, South Carolina, earthquake. It was judged that the selection of 2 per cent in 50 years likelihood as the maximum considered earthquake ground motion would result in acceptable levels of seismic safety for the nation.

The acceleration experienced by a building will vary depending on the period of the building, and in general short-period buildings will experience more accelerations than long-period buildings, as shown in the response spectrum discussed in section 4.5.3. The USGS maps recognize this phenomenon by providing acceleration values for periods

NATURE OF EARTHQUAKES AND SEISMIC HAZARDS

of 0.2 seconds (short) and 1.0 seconds (long). These are referred to as spectral acceleration (SA), and the values are approximately what are experienced by a building (as distinct from the peak acceleration which is experienced at the ground). The spectral acceleration is usually considerably more than the peak ground accelerations, for reasons explained in Section 4.7.

Figure 2-14 shows 2%/50 year probability maps for the central and southern United States for 0.2 seconds, and Figure 2-15 shows a similar map for 1.0 second spectral acceleration.

These USGS probability maps provide the basis for USGS maps used in building codes that provide design values for spectral acceleration used by structural engineers to calculate the seismic forces on a structure. These design value maps differ by use of a maximum considered earthquake (MCE) for the regions. For most regions of the country the maximum considered earthquake is defined as ground motion with a uniform likelihood of exceedance of 2% in 50 years (a return period of about 2,500 years) and is identical to the USGS probability maps. However, in regions of high seismicity, such as coastal California, the seismic hazard is typically controlled by large-magnitude events occurring on a limited number of well-defined fault systems. For these regions, rather than using the 2% in 50-year likelihood, it is considered more appropriate to directly determine the MCE ground motions based on the characteristic earthquakes of those defined faults.

The 2000 *NEHRP Provisions* and the 2003 *IBC* provide maps that show the MCE for the Coterminous United States, California and Hawaii, the Utah region, Alaska, the Puerto Rico region and Guam. These maps are produced in black and white line with no color coding. A CD-ROM is available from FEMA, USGS, ACSF and IBC that includes a software package that can provide map values based on latitude/longitute or postal zip code.

Finally, the acceleration values shown on the maps are not used directly for design. Instead, they are reduced by 1/3; this value is termed the Design Earthquake (DE) and is the value used by engineers for design. The reason for this is that engineers believe that the design provisions contain at least a margin of 1.5 against structural failure. MCE is inferred to

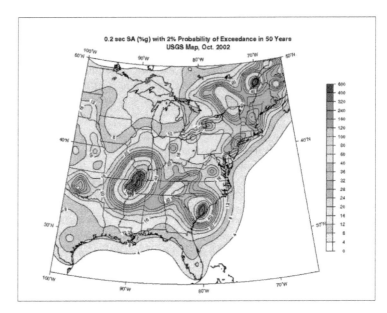

Figure 2-14: 0.2 second (short) period Spectral Acceleration values in %g with a 2% Probability of Exceedance in 50 years for Central and Southern United States.

SOURCE: USGS NATIONAL SEISMIC HAZARD MAPPING PROJECT

provide collapse prevention level, while the actual design is done using the design earthquake (DE), which is 2/3 MCE for code-level life-safety protection-level. This belief is the result of the study of the performance of many types of buildings in earthquakes, mostly in California.

There have been numerous comments that the level of seismic hazard being used in the central and eastern United States results in design values that are unreasonably high. As a result, a review and re-verification of the 2% in 50 years ground-shaking probability for use as the MCE will be implimented. This study is being done as part of the 2008 *NEHRP Recommended Provisions* update process.

## 2.6.3 Rapid Response: ShakeMaps

An important point in summarizing the present status of assessment of seismic strong ground motions is that in a number of countries, digital

NATURE OF EARTHQUAKES AND SEISMIC HAZARDS

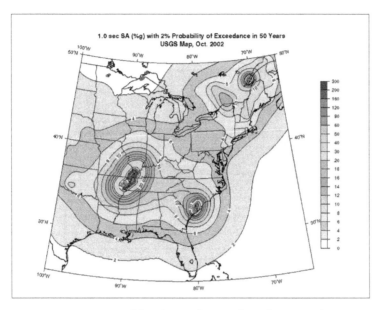

Figure 2-15: 1.0 second (long) period spectral acceleration values in %g with a 2% probability of exceedance in 50 years for central and southern United States.

SOURCE: USGS NATIONAL SEISMIC HAZARD MAPPING PROJECT.

strong-motion systems linked to communication centers (telephone, wireless, or satellite) have now been installed. These provide processed observational data within a few minutes after shaking occurs. The USGS **ShakeMap** program produces a computer-generated representation of ground shaking produced by an earthquake. (Figure 2-16) The computation produces a range of ground shaking levels at sites throughout the region using attenuation relations that depend on distance from the earthquake source, and the rock and soil conditions through the region so that the observed strong ground motions can be interpolated. One format of the maps contours peak ground velocity and spectral acceleration at 0.3, 1.0, and 3.0 seconds and displays the locational variability of these ground motion parameters.

Not only peak ground acceleration and velocity maps are computed using instrumental measurements, but by empirical correlations of the various scales, approximate Modified Mercalli Intensity estimates are also

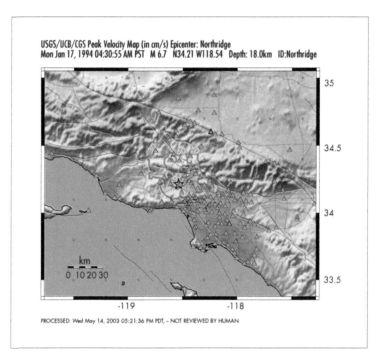

USGS/UCB/CGS Peak Velocity Map (in cm/s) Epicenter: Northridge
Mon Jan 17, 1994 04:30:55 AM PST   M 6.7   N34.21 W118.54   Depth: 18.0km   ID:Northridge

PROCESSED: Wed May 14, 2003 05:21:36 PM PDT, – NOT REVIEWED BY HUMAN

Figure 2-16: Example of a peak ground velocity (PGV) ShakeMap for the 1994 Mw=6.7 Northridge earthquake.

Strong-motion stations are shown as triangles, the epicenter as a red star, and thick red lines show contours (30, 60, and 90 cm/s) of PGV. Directivity during the rupture process causes the largest amplitudes to be located significantly to the north of the epicenter.

mapped. These maps make it easier to relate the recorded ground motions to the felt shaking and damage distribution. In a scheme used in the Los Angeles basin, the Instrumental Intensity map is based on a combined regression of recorded peak acceleration and velocity amplitudes (see Wald et al., 1999).

In 2001, such ShakeMaps for rapid-response purposes became available publicly on the Internet (see Section 2.10) for significant earthquakes in the Los Angeles region and the San Francisco Bay Area of California. Similar maps are available in other countries. Additionally, efforts are underway to combine near-real-time knowledge about the earthquake source process with the observed strong ground motions to produce maps that may better take into account the effects due to directivity.

NATURE OF EARTHQUAKES AND SEISMIC HAZARDS

ShakeMaps represent a major advance not only for emergency response, but also for scientific and engineering purposes. Their evolution and improvement will no doubt be rapid.

## 2.7 CONCLUSIONS

The seismological methods dealt with in this chapter will no doubt be much extended in subsequent years. First, greater sampling of strong-ground motions at all distances from fault sources of various mechanisms and magnitudes will inevitably become available. An excellent example of seismic recording growth comes from the 1999 Chi-Chi, Taiwan, earthquake.

Another interesting recent case is the major Alaska earthquake of November 3, 2002. This 7.9 magnitude earthquake was caused by rupture along the Denali fault for 200 km, with right-lateral offsets up to 10 m. A number of strong-motion records were obtained; the Trans-Alaskan oil pipeline did not suffer damage because of an innovative pipeline design combined with sophisticated knowledge of the seismology.

Second, more realistic 3D numerical models will solve the problem of the sequential development of the wave mixtures as the waves pass through different geological structures. Two difficulties may persist: the lack of knowledge of the roughness distribution along the rebounding fault and, in many places, the lack of quantitative knowledge of the soil, alluvium, and crustal rock variations in the region. For these reasons, probabilistic estimation as a basis of engineering decisions seems preferable.

Over the past decade, advances in digital seismometry have greatly reduced the recovery and computer processing time of recorded data, producing near-real time analysis products important for post-earthquake emergency response (Gee et al., 1996; Dreger and Kaverina, 2000; Wald et al., 1999). Continuing improvements in technology are expected to further increase the amount of timely earthquake source and strong ground motion information. A recent significant advance in general motion measurement is correlation with precisely mapped co-seismic ground deformations, and efforts are currently underway to obtain and analyze these data in near-realtime. Networks of continuous, high-sample-rate **Global-Positioning-System** (GPS) instruments will no doubt help greatly in future understanding of the source problem and the correct adjustment to strong-motion displacement records.

A broad collection of standardized strong-motion time histories is now being accumulated in virtual libraries for easy access on the Internet. Such records will provide greater confidence in seismologically sound selection of ground motion estimates.

Additional instrumentation to record strong ground motion remains a crucial need in earthquake countries around the world. Such basic systems should measure not only free-field surface motions, but also downhole motions to record the wave changes as they emerge at the earth's surface.

The Advanced National Seismic System (ANSS) program is a major USGS and NEHRP initiative that provides accurate and timely information on seismic events. It is working to unify seismic monitoring in the United States, and provides a framework to modernize instrumentation and revolutionize data availability for research, engineering and public safety. (For more information, see http:www.anss.org/.)

In particular, many contemporary attenuation estimates for ground velocity and displacement will no doubt be improved as more recorded measurements are included, rendering earlier models obsolete. The statistical basis for separation of the probability distributions as functions of the various key parameters will become more robust. To keep abreast of changes, ground motion attenuation model information may be found at the USGS Earthquake Hazards Program website (see Section 2.10).

## 2.8 ACKNOWLEDGMENTS

We thank the personnel of the U. S. Geological Survey for their efforts in preparing USGS Fact Sheet 017-03, from which several figures were obtained.

## 2.9 CITED AND OTHER RECOMMENDED REFERENCES

Abrahamson, N. A. and W. Silva, 1997, Empirical response spectral attenuation relations for shallow crustal earthquakes, *Seism. Res. Letters*, 68, 94-127.

Atkinson, G. M., and D. M. Boore, 1995, New ground motion relations for eastern North America, *Bull. Seism. Soc. Am.*, 85, 17-30.

Bolt, B. A., 1975, The San Fernando earthquake, 1971. *Magnitudes, aftershocks, and fault dynamics*, Chap. 21, Bull. 196, Calif. Div. of Mines and Geol., Sacramento, Calif.

Bolt, B. A. 2003. Engineering seismology, In: *Earthquake Engineering, Recent Advances and Applications*, eds. Y. Bozorgnia and V. V. Bertero, CRC press. Florida.

Bolt, B. A., 2003, *Earthquakes*, 5th edition, W. H. Freeman: New York.

Comerio, M., 1998. *Disaster Hits Home*, University of California Press: Berkeley.

FEMA 450, NEHRP (National Earthquake Hazards Reduction Program) *Recommended Provisions and Commentary for Seismic Regulations for New Buildings and Other Structures*, 2003 Edition, Federal Emergency Management Agency, Washington, DC

Gee, L. S., D. S. Neuhauser, D. S. Dreger, M. Pasyanos, R. A. Uhrhammer, and B. Romanowicz, 1996, Real-time seismology at UC Berkeley: The Rapid Earthquake Data Integration Project, *Bull. Seism. Soc. Am.*, 86, 936-945.

Hanks, T C, and H. Kanamori, 1979. A moment magnitude scale, *Journ. Geophys. Res.*, 84, 2348-2350.

International Code Council, *International Building Code* (IBC), 2003, Falls Church, VA

National Fire Protection Association, NFPA 5000, *Building Construction and Safety Code* , 2006, , Quincy,MA

Olsen, K.B., R.J. Archuleta, and J.R. Matarese (1995). Three-dimensional simulation of a magnitude 7.75 earthquake on the San Andreas fault, *Science*, 270, 1628-1632.

Stein, R. S., 2003, Earthquake Conversations, *Scientific American*, Vol. 288, No. 1, 72-79.

Yeats, R. S., C. R. Allan, K. E. Sieh, 1997. *The Geology of Earthquakes*, Oxford University Press.

Yeats, R. S., 2001. *Living with Earthquakes in California: A Survivor's Guide*, Oregon State University Press: Corvallis.

## 2.10 WEB RESOURCES

Consortium of Organizations for Strong-Motion Observational Systems COSMOS  http://www.cosmos-eq.org

European Strong-Motion Database (ISESD)
http://www.isesd.cv.ic.ac.uk/

National Seismic Hazard Mapping Project, Golden, Colorado
http://geohazards.cr.usgs.gov/eq/

ShakeMaps  www.trinet.org/shake

Tsunami Warning Centers http://www.prh.noaa.gov/pr/ptwc/
http://wcatwc.gov/

USGS Earthquake Hazards Program  http://earthquake.usgs.gov/

by Richard Eisner

## 3.1 INTRODUCTION

This chapter describes how earthquake hazards can affect site selection
and planning, and the process for identification of site and regional
factors that impact seismic design. Site selection is typically determined
by initial land costs, land use criteria such as zoning, proximity to transportation, and utility infrastructure. Additional site location factors
that should be considered include environmental and geotechnical site
conditions that would impact building performance, and factors that
influence structural design criteria that would impact costs and performance.

The importance of a design team comprised of the client, architect, the
geotechnical civil engineer and structural engineer is emphasized, and
a process for geotechnical assessment of a site is identified. Regional
factors of earthquake probabilities and ground motions are identified
and reviewed at the project level. The interaction of the regional risk,
building program, and client expectations is discussed in the context of
performance objectives. Site hazards are identified, and mitigation approaches are presented.

## 3.2 SELECTING AND ASSESSING BUILDING SITES
## IN EARTHQUAKE COUNTRY

In earthquake hazard areas, selection and evaluation of the site will be
critical to meeting client expectations on project performance. Identification and analysis of the threat posed by earthquakes to a specific
location or site are more complex and frequently less precise than
analysis for hazards such as flood or wind, where information about
frequency and intensity of events is well documented. For example,
the threat of flood is defined by 100-year flood zones delineated by the
National Flood Insurance Program[1] (1% probability of being exceeded
per year) and is mapped at the parcel level. A site is either in or out of
the flood zone. Areas at risk to earthquake damage encompass entire
regions, not just the areas adjacent to faults. The zones of potential
damage are not neatly defined or delineated. There are numerous fac-

---

[1] notes will be found at the end of the chapter

tors in addition to shaking that will affect a building's performance and its continued function. Therefore, at the onset of a project, a thorough examination should be undertaken of regional potential for earthquakes and the areas that will be damaged by ground faulting, ground shaking, subsidence and liquefaction, utility disruption and from the secondary hazards and impacts from earthquake-caused fires, floods, and hazardous materials releases.

### 3.2.1 Performance Criteria, Site Selection, and Evaluation

Site selection criteria should be derived from the building's program and performance-based design criteria. A simple project may only be designed to the minimum level of performance - life safety. Such a structure is only expected to protect the lives of occupants and may be so extensively damaged after a quake that it will have to be demolished. With a small project where client performance criteria are limited, the site evaluation criteria will be focused on the immediate site environment, on-site hazards, and adjacent structures and land uses. Geotechnical investigations focus primarily on the site. Mitigation is usually accomplished by providing a setback to separate new construction from adjacent hazards and through design of the foundations and structure to meet the building code. Where the client has higher expectations of building performance, such as minimizing damage and maintaining business operations, the assessment will need to be more rigorous, and the scope of the site investigations will extend far beyond the "property line" to include all of the potential hazards that would influence continuity of operations, including land uses in proximity to the site and area access and egress, utility performance, the need for alternative lifeline capability (back-up generators, water and waste water processing and storage, alternate telecommunications, etc.).

For facilities designed to performance-based criteria, including minimum disruption and continued operation, the location of the site within the region may play a critical role in meeting client expectations. The definition of "site" becomes the region within which the facility is located, and "vulnerability assessments" must examine both facilities and the connections the facilities have to raw materials, personnel and distribution to markets. This is a more holistic view of building design and vulnerability, that addresses disruption of operations and the economic impacts of disasters.

### 3.2.2 Building Program and Site Evaluation

The development of the program for the building and the definition of performance criteria are iterative processes that take into account the needs of the client, the characteristics of the earthquake hazard, the characteristics of the site (or alternate sites) and availability and cost of engineering solutions to mitigate the hazard (Figure 3-1). If the client wishes the building to withstand a major earthquake without damage and be able to maintain operations, the program will establish both performance and site selection criteria to achieve their goal. The program will also establish utility, access, and egress performance expectations that will influence location.

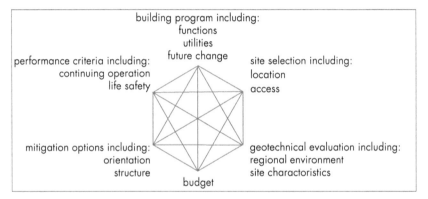

Figure 3-1: Interrelationships of performance expectations, building program and site characteristics.

## 3.3 THE IMPORTANCE OF THE RIGHT TEAM— GEOTECHNICAL ENGINEERING EXPERTISE

Understanding and incorporating the earthquake threat and its impact on a location or facility is a complex assessment process requiring an understanding of the earthquake hazard, how a site will respond to arriving ground motions, and how a structure will interact with the site's motions.

It is therefore essential that the client, architect and structural engineer retain the services of a Geotechnical Engineer to provide input to the assessment of alternate sites, and to assist in the structural design of the programmed facility.

### 3.3.1 The Site Assessment Process

The California Geological Survey (CGS) provides guidance in the use of geotechnical and civil engineering expertise in *Guidelines for Evaluating and Mitigating Seismic Hazards in California.*[2] The Guideline emphasizes the need for both geotechnical engineering to identify and quantify the hazard, and civil engineering to develop mitigation options for the architect and owner. Chapter 3 of the Guideline provides recommended site investigations for assessing seismic hazards and is summarized below.

### 3.3.2 Geotechnical Report Content

The geotechnical investigation of the site is a vital resource to designer and structural engineer in designing and building an earthquake-resistant structure. The CGS recommends that a geotechnical report include the following data:

○ Description of the proposed project location, topography, drainage, geology, and proposed grading.

○ Site plan indicating locations of all tests.

○ Description of the "seismic setting", historic seismicity and location of closest seismic records used in site evaluation.

○ Detail (1:24,000) geologic map of the site indicating pertinent geologic features on and adjacent to the site.

○ Logs of all boring or other subsurface investigations.

○ Geologic cross section of the site.

○ Laboratory test results indicating pertinent geological data.

○ Specific recommendations for site and structural design mitigation alternatives necessary to reduce known and/or anticipated geologic and seismic hazards.

### 3.3.3 Additional Investigations to Determine Landslide and Liquefaction

Additional tests may be necessary to determine if there is a potential for earthquake induced landslides and/or liquefaction. These tests and procedures are identified in Recommendation Procedures for Implementation of California Department of Mines and Geology (CDMG) Special Publication 117: *Guidelines for Analyzing and Mitigating Liquefaction in California;* and *Recommended Procedures for Implementation of DMG Special Publication 117: Guidelines for Analyzing and Mitigating Landslide Hazards in California.*[3]

### 3.3.4 Information Sources for the Site Assessment Process

In evaluating or selecting a site, the objective will be to identify those natural and man-made forces that will impact the structure, and then to design a site plan and the structure to avoid or withstand those forces. It is necessary to start the site evaluation process with research of information available from local building and planning departments, the National Weather Service, FEMA's National Flood Insurance Program (NFIP), the United States Geological Survey, state geological surveys, university geology departments and published research, and a geotechnical engineering firm familiar with the region and sources of local information. Include, where available, hazard mapping zones of ground faulting, liquefaction, landslides and probabilistic assessments of ground motions.

Where sites are within a mapped hazard zone, a site-specific investigation should be conducted by a geotechnical engineer to identify or demonstrate the absence of faulting, liquefaction or landslide hazards. When a hazard is identified and quantified, recommendations for mitigation should be provided. The following information will assist in assessing the geotechnical hazards in a region or on a site:

○ Topographic, geologic and soil engineering maps and reports, and aerial photographs

○ Water well logs and agricultural soils maps.

○ State hazard evaluations maps.

FEMA's *Understanding Your Risks: Identifying Hazards and Estimating Losses* (FEMA 386-2)[4] provides an excellent example of a hazard assessment process that can be adapted to your practice.

When a site is outside a mapped hazard zone, ensure that proposed development and alterations to the site do not increase susceptibility to hazards (such as cuts and fills that increase ground water percolation or increase the likelihood of earthquake-induced landslides).

## 3.4 LOCAL GOVERNMENT HAZARD ASSESSMENTS—DMA 2000

In 2000, Congress amended the Stafford Act (federal legislation that provides pre- and post-disaster relief to local and state governments), adding requirements that local governments, states and tribes identify and develop mitigation plans to reduce losses from natural hazards. The Disaster Mitigation Act of 2000 (DMA 2000)[5] requires these governments to identify and map all natural hazards that could affect their jurisdictions. Beginning in November 2004, local and state governments were to be able to provide an architect or engineer with hazard and risk assessments for earthquakes, flooding, landslides, tsunami, and coastal erosion. The risk assessments[6] are intended to be the basis for land use development decisions and for setting priorities for local and federal mitigation funding, but they will also provide a basis for initial site selection and evaluation.

## 3.5 TOOLS FOR GETTING STARTED

As noted in Chapters 2 (Section 2.6.2) and 4, earthquakes produce complex forces, motions and impacts on structures. Between the earthquake and the structure is the site, which determines how the building experiences the earthquake, and what secondary hazards are triggered by ground motions. These additional hazards include surface rupture or faulting; near-source effects of strong ground motions; ground failure and landslides, subsidence; and lateral spreading and liquefaction. In coastal regions, in areas within dam inundation zones, or areas protected by earth levies, flooding can occur as a result of dam or levy failure triggered by ground motions, or, in coastal areas, by earthquake-triggered tsunamis. Each of these primary and secondary hazards should be identified in the site assessment and mitigated where they would adversely impact building performance. The following sections will elaborate on each of these site hazards and identify mitigation alternatives.

## 3.5.1 Understanding Regional Earthquake Risk- Big Picture of Expected Ground Motions

There are a number of resources available that provide a regional view of the earthquake hazard. Overall assessments of risk are expressed as probabilities that mapped ground motions will exceed a certain level over a period of time. A common measure is the 10% probability that peak ground acceleration (violence of ground shaking) will be larger than the value mapped, over a 50-year period. These maps provide an assessment of the relative intensity of ground motions for a region.

● USGS 2002 Ground Motion Maps

The building code uses the Maximum Considered Earthquake (MCE) maps which are based on the USGS Seismic Hazard Maps with a 2% probability of being exceeded in about 2,500 years (Figure 3-2). See section 2.6.2. These maps depict areas that have an annual probability of approximately 1 in 2,300 of the indicated peak ground acceleration being exceeded, and account for most known seismic sources and geological effects on ground motions. The areas of intense orange, red, brown and black are the most likely areas to experience violent ground shaking greater than 30% of the force of gravity in the next 50 years. The maps provide a general assessment of relative ground motions, but are at a scale that does not help in a site selection process. It is clear from the map, however, that violent ground motions are more likely in the coastal regions of California, Oregon and Washington, the Sierra Nevada range of California, and the Wasatch Range of Utah than in Colorado, Kansas and Oklahoma.

● State Survey Risk Maps

Many states provide geological data that can assist in assessing regional seismic risk. In California, for example, the CGS in cooperation with the USGS has taken the data from the above map and provided a more detailed set of regional maps. The map of the Bay Area (Figure 3-3) depicts the peak ground accelerations with a 2% probability of being exceeded in 50 years at a regional scale and combines probability of occurrence of large ground motions and soil and geological conditions that would amplify ground motions. The areas depicted in red through gray are the

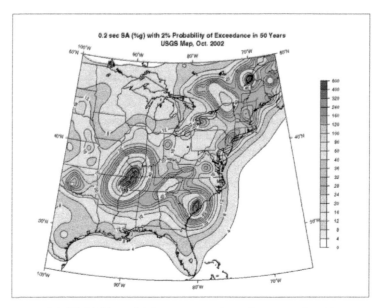

Figure 3-2: USGS Map of 0.2 sec spectral acceleration with a 2%
Probability of Exceedance in 50 Years

SOURCE: USGS §201.6(C)(2), 44 CFR PART 201, STATE AND LOCAL PLAN INTERIM CRITERIA
UNDER THE DISASTER MITIGATION ACT OF 2000 [7]

areas where the most violent ground shaking will most frequently occur.
These areas are adjacent to active faults capable of producing violent
ground motions and areas where soils conditions will increase ground
motions. Thus, areas in gray are adjacent to active faults and along the
margins of the San Francisco Bay, where unconsolidated soils will amplify
ground motions. It is important to note, however, that the map depicts
probability of relative shaking and that damaging ground motions can
occur anywhere in the region depicted on the map.

Both the USGS and CGS depict the ground motions that are produced
by all earthquakes on all faults that could influence a particular location.
These maps can be extremely helpful to the architect and client in determining the relative risk of alternative sites and the trade-offs of location,
vulnerability and offsetting costs for a structure that will resist ground
motions.

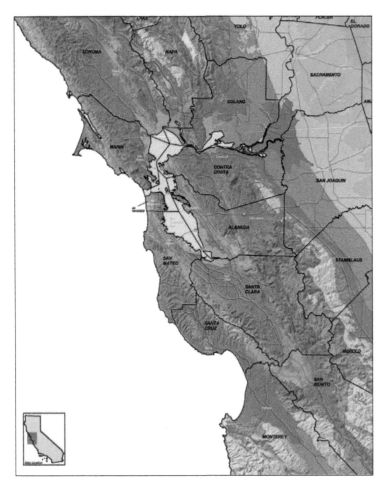

Figure 3-3: USGS and CGS Map of Relative PGA with a 2% Probability
of Being Exceeded in 50 Years

SOURCE: CALIFORNIA SEISMIC SAFETY COMMISSION, USGS, CGS AND OES, 2003 [8]

● HAZUS Earthquake Loss Estimates[9]

The Federal Emergency Management Agency (FEMA) has developed a
software program that can be used to estimate earthquake damage and
losses at a regional level. HAZUS (Hazards United States) provides estimates of damage and losses to infrastructure such as highway bridges,
electrical and water utilities, casualties and requirements for shelter

**Figure 3-4**

HAZUS Estimate of Peak Ground Acceleration (PGA) by Census Tract, Alemeda County, California.

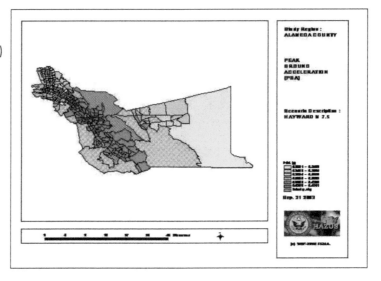

for displaced households. The quality of a HAZUS estimate of losses will depend on the detail of information input to the program. Local soil data and building inventory will determine the accuracy of the loss estimates. Estimates can be produced for specific faults, for specific scenarios, or for annualized losses over a period of years. In each case, the loss estimate is helpful in understanding the "risk context" for a project -what damage and disruption will occur in the community surrounding the project. Below are two HAZUS maps (Figures 3-4 and 3-5) illustrating intensity of ground motions in PGA and Total Economic Losses, by census tract for a M7.5 earthquake on the Hayward Fault in Alameda County, California. Similar estimates can be produced for other areas of the country where an earthquake threat would influence site selection. While HAZUS is helpful in understanding regional vulnerability and patterns of damage and loss, it is not appropriate for assessment of damage on an individual building site.

Information about HAZUS is available from your local and state emergency services office, and from FEMA at www.fema.gov/hazus/hz_index.shtm and www.hazus.org.

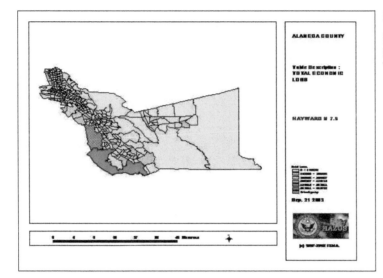

Figure 3-5

HAZUS Estimate of Total Economic Loss by Census Tract

## 3.6 EARTHQUAKE HAZARDS TO AVOID

The most obvious manifestations of earthquakes are earthquake fault offset, liquefaction, landslides, and ground shaking (Figures 3-6 and 3-9). Each of these hazards can be mitigated through careful site planning. Examples in this section are drawn from California, where a broad range of hazard identification and mitigation approaches is available. Hazard data and land regulation practices vary from state to state and within states. A geotechnical consultant is your best source of local data.

### 3.6.1 Earthquake Fault Zones

The United States Geological Survey and many state geological surveys produce maps of active earthquake faults - that is, faults that exhibit "Holocene surface displacement" or ruptured within the last 11,000 years. These maps depict faults where they have ruptured the ground surface, as fault movements usually recur in geologically weak zones. In California, the legislature mandated the mapping of active faults after the 1971 San Fernando earthquake (Alquist-Priolo Earthquake Fault Zoning Act)[10]. The Fault Hazard Zones Act maps are published by the state, and location of a site within a Fault Hazard Zone requires disclosure of the hazard at point of sale. Local governments are responsible for reviewing

Figure 3-6

Landers-Big Bear Earthquake (1992). Ground faulting extended for nearly 50 miles.

geologic reports and approving those reports before approving a project (Figure 3-7).

Fault mapping is a continuing process of discovery, analysis, and mapping. However, it is important to note that not all earthquake faults rupture to the surface and not all earthquake faults are currently mapped. In the 1989 Loma Prieta earthquake (M6.7), the fault did not rupture to the surface, yet caused more than $6 billion in damage and resulted in more than 60 deaths. For some active faults, there may not be a surface manifestation indicating recent activity. Both the 1971 San Fernando and 1994 Northridge earthquakes occurred on blind thrust faults, where faulting did not reach the surface, so the hazard was not recognized until the earthquake occurred. Nonetheless, fault zones pose a clear danger to structures and lifelines, and where formally mapped or inferred from geologic reports, site plans should provide a setback to protect structures from fault movement.

● Mitigating Fault Zone Hazards

The Alquist-Priolo Earthquake Fault Zoning Act[12] provides reasonable guidance that should be applied in site selection and site design. Requirements include:

○ Disclosure that a property is within a mapped Seismic Hazard Zone. The zones vary in width, but are generally ¼ of a mile wide and are defined by "turning points" identified on the zone maps. Figure 3-8

Figure 3-7

Principal Active Fault Zones in California

SOURCE: CGS, SPECIAL PUBLICATION 1999

shows a zone map that identifies active traces of the fault, the date of last rupture, and defines the "fault zone" within which special studies are required prior to development. The zone boundary, defined by turning points, encompasses known active traces of the fault and provides approximately 200 meters setback between the fault trace and the boundary line. Check with local government planning agencies for the most current maps.

○ Local governments must require a geologic report for any project within the fault hazard zone to ensure that no structure is built across an active fault trace.

○ No structures for human habitation shall be built within 50 feet of an identified fault trace (an exception is provided for single-family residential structures when part of a development of four or fewer structures).

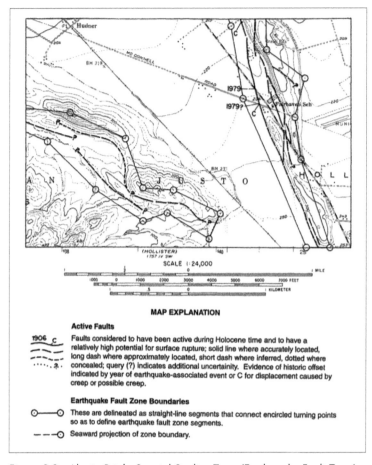

Figure 3-8: Alquist Priola Special Studies Zone (Earthquake Fault Zone).

SOURCE: CGS 1999, *FAULT - RUPTURE HAZARD ZONES IN CALIFORNIA*, SPECIAL PUBLICATION 42.
AVAILABLE AT FTP://FTP.CONSRV.CA.GOV/PUB/DMG/PUBS/SP/SP42.PDF

In states where seismic hazard zones are not identified, the geologic report for a project should locate identified or suspected fault traces, and recommend mitigation measures, including those identified above, to reduce the risk posed

SITE EVALUATION AND SELECTION

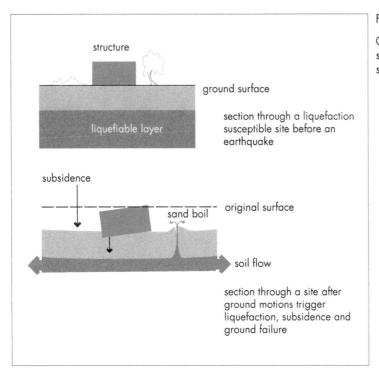

Figure 3-9

Cross section through a site where liquefaction and subsidence could occur.

structure

ground surface

section through a liquefaction susceptible site before an earthquake

liquefiable layer

subsidence

original surface

sand boil

soil flow

section through a site after ground motions trigger liquefaction, subsidence and ground failure

## 3.6.2 Ground Failure Due to Liquefaction

Liquefaction occurs when water-saturated soils, sands, or gravels flow laterally or vertically like a liquid. This occurs when earthquake ground motions shake the material until the water pressure increases to the point that friction between particles is lost, and the ground flows, losing its strength (Figure 3-9). Liquefaction is most likely to occur where the soils are not consolidated (near rivers and streams, in basins, near coast-lines and in areas of unconsolidated alluvium) and where ground water is within three to four meters of the surface. Liquefaction can occur at greater depths, resulting in large-scale ground failure that can destroy pavement, underground utilities, and building foundations (Figure 3-10). The subsidence of Turnagain Heights in Anchorage during the 1964 earthquake is an example of deep-seated liquefaction and ground failure. When a soil liquefies, it can flow laterally, eject vertically as a sand boil, or result in subsidence and ground failure (Figure 3-11).

Figure 3-10

Liquefaction in San Francisco's Marina District (1989)

SOURCE: USGS

Sand boils and flows on the surface can displace and damage structures and utilities. Lateral liquefaction flows will result in subsidence, loss of foundation integrity, disruption of underground utilities and damage to structures resting on the soil surface, including roadways and utility structures. Liquefaction susceptibility and potential should be identified in the site geotechnical investigation, as explained in Section 3.3.3.

● Liquefaction Hazard Zones

Liquefaction susceptibility can be determined from site geologic investigations and from a review of geologic and soil maps and water well and bore hole logs. In California, liquefaction potential mapping is part of

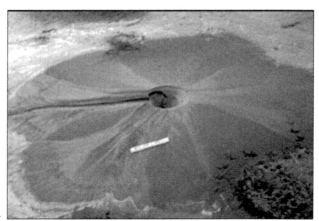

Figure 3-11: Sand Boil.

the CGS's Earthquake Hazard Mapping Program. Liquefaction hazard zone maps have been completed for sections of the Los Angeles and San Francisco Bay Regions (Figures 3-12 and 3-13).

Within an identified liquefaction hazard zone, maps of liquefiable soils, prepared by a geotechnical engineer, should identify the location and extent of "cohesionless silt, sand, and fine-grained gravel in areas where the ground water is within 50 feet of the surface." Procedures for testing and criteria for determining liquefaction susceptibility are contained in *Recommended Procedures for Implementation of DMG Special Publication 117: Guidelines for Analyzing and Mitigating Liquefaction in California.*[13]

● Mitigation Options for Liquefiable Sites

There are structural solutions for mitigating liquefaction potential that address the design of foundation systems that penetrate the liquefiable layers. It should be noted that while it is frequently cost effective to design structures to withstand liquefaction, making access and egress routes, parking and storage facilities and above and underground utilities "liquefaction-resistant" is prohibitively expensive. "A whole-site solution" may be more practical when site choice is limited and susceptibility is significant. See the mitigation approaches below.

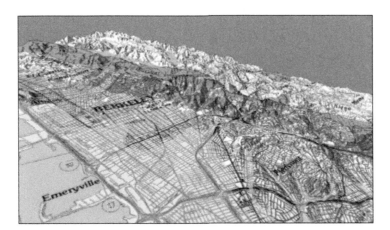

Figure 3-12: 3-D image of Liquefaction and Landslide Hazard Zone Map for Berkeley and Emeryville. Yellow indicates liquefaction, which is related to soil type and proximity to ground water.

SOURCE: CGS

Figure 3-13: Liquefaction Hazard Zone Map for West Oakland and Emeryville.

SOURCE: CGS EARTHQUAKE HAZARD MAPPING PROGRAM

● Location of the Structure

The simplest way to mitigate the potential of liquefaction is to avoid those locations in a region or on a site where the potential for ground failure is identified in the geotechnical investigation. Locate structures where ground water is low, where soils are compacted, and where soils are not homogeneous sands or gravels.

● Intervention on the Site

While avoidance is the optimum solution, it is not always possible. Mitigating liquefaction potential involves changing the characteristics of the site. The following options are all costly and vary in extent of risk mitigation. Seek advice from geotechnical and civil engineering consultants about the most cost-effective intervention.

SITE EVALUATION AND SELECTION

○ Site Compaction

On sites with unconsolidated soils, the response of the site can be improved by compacting the soil, compressing it so that soil particles are forced together, reducing water-filled voids and increasing the friction between soil particles.

○ Change Soil

The performance of the site can also be improved by excavation of the liquefiable soils and replacement with compacted hetero-geneous fill. By changing the soil, the susceptibility of the site to liquefaction will be significantly reduced. However, for both this approach and the compaction alternative, site performance is improved by construction of barriers to the infiltration of water so that the groundwater level of the site is lowered.

○ Dewatering the Site

An alternative to "reconstituting the site" by replacing the soil is to dewater the site. This approach requires constructing wells to pump out and lower the ground water level to reduce liquefaction susceptibility. To reduce the demand for continuous pumping, dewatering should be combined with the construction of infiltration barriers. A back-up power source to ensure post-disaster pump operations should be provided.

● Special Design Considerations

As noted above, the potential for liquefaction of a site poses severe problems for maintenance of access and egress and performance of lifelines including power, telecommunications, water sewer and roadways. For facilities that are expected to be in continuous operation after disasters, redundant access to utility networks, multiple access and egress paths, and back-up power and communication systems should be provided. Liquefaction potential may be difficult to assess, so a conservative approach to the design of continuous operation facilities is essential.

## 3.6.3 Areas of Intensified Ground Motions

Local geology, proximity to faults and soil conditions play significant roles in how earthquake forces impact a structure. The Loma Prieta earthquake (1989) provided a striking example of how local soils and regional geology can determine damage. Sixty miles from the earthquake's epicenter in the Santa Cruz Mountains, the soils determined the pattern of damage to the Cypress Viaduct in Oakland. As illustrated below, the damage corresponded to the quality of the ground. On bedrock materials in the East Bay hills of Oakland, ground motions were small and there was little damage. On sandy and gravel soils between the East Bay hills and the San Francisco Bay, the amplitude of ground motions increased, but there were few collapsed structures. However, on the soft mud adjacent to the Bay, the amplitude of the ground motions and the duration of strong shaking increased. The Cypress Structure, where it passed from "sand and gravel" to "soft mud" collapsed (Figure 3-14).

A similar condition existed in the Marina District of San Francisco where soft soils liquefied, amplified motions and extended the duration of shaking until several structures collapsed, while elsewhere in San Francisco, on firmer ground, there was little or no damage.

Figure 3-15 was developed by the USGS and predicts amplification of ground motions based on soil types adjacent to San Francisco Bay. According to the USGS, "this map shows the capability of the ground to amplify earthquake shaking in the communities of Alameda, Berkeley, Emeryville, Oakland, and Piedmont. The National Earthquake Hazards Reduction Program recognizes five categories of soil types and assigns amplification factors to each. Type E soils in general have the greatest

Figure 3-14

Comparison of ground motions under the Cypress Viaduct, Loma Prieta Earthquake 1989.

SOURCE: GRAPHIC FROM THE USGS

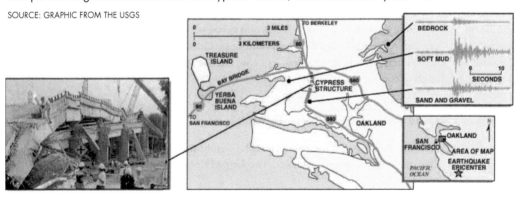

SITE EVALUATION AND SELECTION

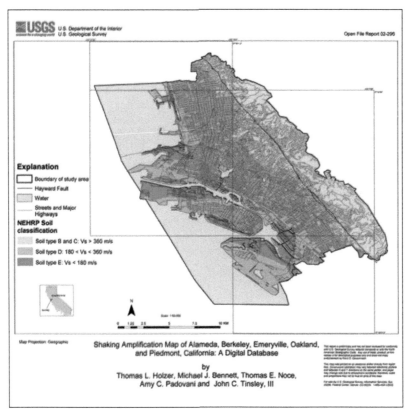

Figure 3-15: USGS Shaking Amplification Map of Alameda, Berkeley, Emeryville, Oakland and Piedmont, California.     SOURCE: USGS (HOLZER ET AL.)

potential for amplification, and type A soils have the least. These soil types are recognized in many local building codes. Records from many earthquakes show that ground conditions immediately beneath a structure affect how hard the structure shakes. For example, sites underlain by soft clayey soils tend to shake more violently than those underlain by rock. The map depicts the amplification potential at a regional scale, and it should not be used for site-specific design. Subsurface conditions can vary abruptly, and borings are required to estimate amplification at a given location."

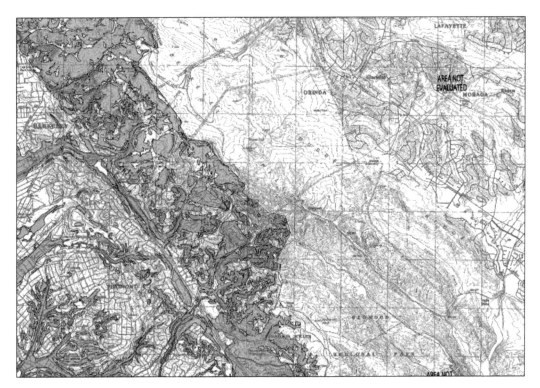

Figure 3-16: Landside Hazard Zones for Berkeley and Oakland, California. Blue areas are those that are susceptible to earthquake-caused landslides.

SOURCE: CGS

### 3.6.4 Ground Failure, Debris Flows, and Land Slides

Potential for ground failure and landslides is determined by soil type, water content (degree of saturation), gradient (slope angle) and triggering events (an earthquake, excavation that upsets the site equilibrium, increase in water content resulting from irrigation or storm run-off). Geotechnical investigations of the site and surrounding terrain are critical in determining site vulnerability.

● Landslide Hazard Maps

The USGS and CGS have prepared Landslide Hazard Zone Maps for parts of northern and southern California (Figure 3-16). The map de-

SITE EVALUATION AND SELECTION

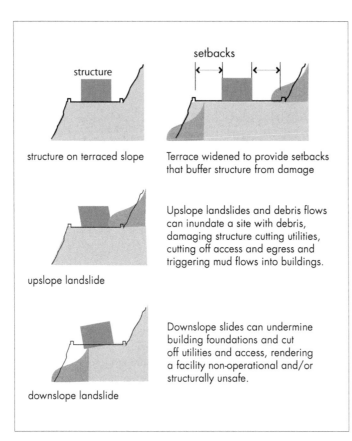

Figure 3-17

The effect of upslope and downslope landslides.

structure

setbacks

structure on terraced slope

Terrace widened to provide setbacks that buffer structure from damage

upslope landslide

Upslope landslides and debris flows can inundate a site with debris, damaging structure cutting utilities, cutting off access and egress and triggering mud flows into buildings.

downslope landslide

Downslope slides can undermine building foundations and cut off utilities and access, rendering a facility non-operational and/or structurally unsafe.

picts a section of the Oakland-Berkeley East Bay Hills, indicating areas where slope, soil type and seismic risk could trigger landslides. Construction in the Landslide Hazard Zone requires an assessment by a geotechnical engineer.

Downslope slides can undermine building foundations and cut off utilities and access, rendering a facility non-operational and/or structurally unsafe (Figure 3-17). The USGS's *National Landslide Hazards Mitigation Strategy*[14] and the California Geological Survey's *Recommended Procedures for Implementation of DMG Special Publication 117: Guidelines for Analyzing and Mitigating Landslide Hazards in California*[15] offer guidance in determining landslide vulnerability and mitigation options.

● Mitigation Options

Foundation systems and structures can be designed to reduce damage from ground failure. The geotechnical engineer can provide recommendations for appropriate foundation design.

○ Set-Back

The most failsafe option for mitigation is to locate structures and lifelines in parts of the site that are not at risk to slide damage. Set back structures from both the toe of an upslope and from the lip of a down slope. Allow separation to accommodate catch basins, debris diverters and barriers. Parking lots or storage areas can be designed and located to "buffer" structures from debris.

○ Drainage

Since water acts as a lubricant on slope-failure surfaces, it is critical that the site and its surroundings be well drained, that irrigation is limited, and that dewatering systems reduce subsurface hydrostatic (water pressure) pressures. Dewatering systems can either be passive (drains into slopes, "French drains," top-of-slope catch basins) or active, providing pumping of subsurface water from sumps into a drain system. In both cases, continuous maintenance is essential to ensure reliable operation of the system. Emergency power may also be required for active drainage systems. Where storm water runoff must be managed on site, design of parking and landscaped areas should accommodate storage. Facility access procedures will need to address displacement of parking and limitations on access during periods the site is flooded.

○ Redundant Infrastructure

Ground failure can severely disrupt utility and lifeline connections to a site. Where continued operations are essential to a client, connections to utility and transportation networks should be redundant, providing more than one means of connection, access, and egress. For telecommunications, redundancy would include dedicated connections to two different switching offices, planned to follow two different routes to the site. Multiple access and egress paths should also be provided. For facilities dependent on electrical power, multiple, dispersed connections to the grid, co-generation and/or emergency back-up power generation should be planned.

Where continuous operation is not essential, emergency back-up power should be provided to ensure safety, security and operation of environmental protection systems (such as heating and ventilating systems [HVAC], water pumps, security systems, evacuation and lighting systems, computer operations and data security. Emergency power generation capacity should exceed minimum requirements to ensure adequate power for projected needs of essential systems. For facilities where a consistent quality-controlled supply of water is essential for operations, on-site storage and purification should be provided to meet operational needs until alternative sources can be secured.

## 3.7 OFF-SITE ISSUES THAT AFFECT SITE SELECTION

As noted previously, for facilities designed to performance-based criteria, including minimum disruption and continued operation, the location of the site within the region may play a critical role in mitigation options. A vulnerability assessment should address issues of access to and egress from the site to regional transportation and communication systems, the robustness of utilities that support the site, and regional earthquake impacts that would affect site operations.

### 3.7.1 Access and Egress

For manufacturing and essential facilities where access and egress are essential for continued operations, siting decisions should address the vulnerability of access roads, freeways, public and private transit and transportation structures upon which business operations will depend. Selection of a site that provides multiple or redundant access and egress is a good idea. This approach will also be essential if facility operations or production is dependant on access by employees, raw materials, and delivery of products, be they manufactured goods or information, to markets. For example, a number of manufacturing firms have relocated their manufacturing from California to other states where product manufacture and delivery would not be disrupted by earthquake damage to buildings, freeway structures, telecommunications, and the dislocation of employees.

### 3.7.2 Infrastructure

We have become more dependent on infrastructure, particularly high-speed telephony for day-to-day business operations. Most businesses are also totally dependent on electrical power from a regional grid, and water and waste-water disposal from offsite utilities. In assessing the vulnerability of a facility expected to be operational immediately after a disaster, the client and designer should assess the reliability of these infrastructure systems and provide for redundancy and back-up systems. For a critical system such as telephony, data telemetry or just Internet access, redundancy should include multiple access or paths to primary utilities. For example, for critical telephony, redundancy would provide multiple paths to different telephone switching offices and satellite communications capability. For electric power, back-up generators and fuel storage (and contracts with suppliers to provide refueling until utility power is restored) to provide for continued operations would be essential. On-site storage for wastewater would provide redundancy to a sewer system that may be damaged by power loss, earthquake damage or flood.

### 3.7.3 Adjacency

Adjacent land uses may pose a threat to the continued operation of the proposed facility. Collapse-hazard structures can spill debris onto the site, damaging structures or blocking access and egress. Hazardous materials released upwind of the site may force evacuation and shut down of operations. Setbacks from adjacent land uses and separation from adjacent structures should be used to protect structures, access and areas of refuge and to protect against pounding. In addition, HVAC systems may require enhanced design to protect building occupants from hazardous materials plumes.

## 3.8 EARTHQUAKE AND TSUNAMI HAZARDS

A tsunami is a rapid rise in coastal sea level caused by offshore earthquakes that displace the ocean bottom, earthquake-triggered or natural submarine landslides and slumps, volcanic eruptions, or very infrequently by meteor strikes. Tsunami waves have a very long wavelength and travel at approximately 500 miles per hour in the open ocean. As they approach shallow waters, their speed and wavelength decreases, and their height increases dramatically.

While coastal storm surge is well documented and understood, the impacts of tsunamis are not as commonly understood. Storm surge produces higher tides and pounding waves over a period of hours. Tsunamis, generated by distant earthquakes on the Pacific Rim, volcanic eruptions or undersea landslides on near-coast continental shelves, or by near-shore earthquakes, can typically cause unpredictable, high and rapidly changing tidal-like inundation from 1 to 30 feet in height above the tide, carrying flood waters and debris inland in, cases up to 100 feet. Tsunami wave arrival is usually, but not always, preceded by extreme tidal recession. The initial wave is usually followed by secondary tsunami waves for periods lasting up to eight hours. These secondary waves can be higher and carry debris from initial inundation, creating a lethal combination of inundation and battering.

Tsunamis are not limited to the Pacific Coast, Hawaii and Alaska. Earthquakes and volcanic eruptions can generate tsunamis along the US southeast and gulf coasts and the Caribbean, with a remote possibility of volcanic and submarine landslides generating a tsunami that could affect the entire Atlantic coastline.

## 3.8.1    Special Considerations for Coastal Area Site Assessment

Coastlines are dynamic. Beaches erode and migrate, and bluffs collapse as part of the natural process in the coastal zone. Earthquakes can accelerate this process. Site plans must address the dynamic nature of the beach-ocean interface, providing setbacks adequate to accommodate inevitable change. Dramatic changes in short periods of time frequently occur as a result of earthquakes, storms and tsunami. Designing structures to resist coastal forces of wind, flood, storm surge, earthquake, tsunami inundation and battering is a complex problem.

● Mitigating Tsunami and Coastal Surge Hazards

FEMA's Coastal Construction Manual (CCM) identifies a process for evaluation of flood hazards in coastal areas that is applicable to earthquake and tsunami forces as well. Alternatives include locating development above the coastal flood zone, orientating structures to reduce the profile presented to wave action, site-planning options for locating structures, parking and landscaping, altering the site and construction of flood protective structures.[16] FEMA suggests the following critical "Do's and Dont's, edited, abridged and adapted from the California Coastal Commission:

○ DO avoid areas that require extensive grading.

○ DON'T rely on engineering solutions to correct poor design and planning.

○ DO identify and avoid or set back from all sensitive, unstable, and prominent land features.

○ DON'T overlook the effects of infrastructure location on the hazard vulnerability of building sites.

○ DO account for all types of erosion (long-term, storm-induced, stream, and inlets).

○ DO incorporate setbacks from identified high-hazard areas.

○ DON'T forget to consider future site and hazard conditions.

○ DO use a multi-hazard approach to planning and design.

○ DON'T assume that engineering and architectural practices can mitigate all hazards.

○ DO involve a team of experts with local knowledge and a variety of expertise in site evaluation and assessment.

● Specific CCM Recommendations for Site Planning

○ Set back structures beyond the code or zoning minimums to provide an extra margin of safety. It is better to be conservative than to have to relocate a structure in the future. If a structure must be located at the minimum setback, it should be designed to be relocated.

○ Set back structures from the lip of coastal bluffs. See the CCM for recommendations.

○ Be aware of multiple hazards. In many coastal states, coastal structures are subjected to potential storm surge, tsunami, coastal erosion, debris flows, fires and earthquakes!

○ Provide setbacks between buildings and erosion or flood control structures to permit maintenance, strengthening and subsequent augmentation.

In site planning, be aware that vegetation and buildings can become "dislodged" and be driven by wind and wave action into structures. Vegetation may serve to stabilize beach areas, but it may not be able to resist tsunamis. Also beware of land forms that may channel inundation into structures.

● Tsunami-Specific Mitigation

In many coastal communities, tsunami inundation is included in the National Flood Insurance Program. Information concerning the potential for tsunami inundation and maps can be obtained from the local or state emergency management office, FEMA or from NOAA. Detailed inundation projections are being prepared for the coastlines of California, Oregon, Washington, Alaska and Hawaii. The potential for tsunami inundation may also exist for some areas in Puerto Rico and the Gulf and southern Atlantic coast states. Unfortunately, the history of tsunami events in the coastal United States is incomplete. The simplest solution is to avoid new construction in areas subject to tsunami inundation (as a surrogate, maps of areas that have historically been inundated by storm surge may be used). For example, the range of projected tsunami inundation for California's open coast is from 33 to 49 feet(10 to 15 meters), with variation in estuaries and bays. In Oregon, Washington, Alaska and Hawaii, wave heights can be greater. The recommendations of the CCM should be followed for construction in areas where erosion, flooding, hurricanes and seismic hazards exist. In developing a site plan in an area with inundation potential, cluster structures in areas with the lowest risk - generally the highest section of the site.

● Resources for Tsunami Mitigation

The National Tsunami Hazard Mitigation Program of NOAA has prepared a number of resource documents to assist architects and planners in mitigating tsunami risk. The guide, *Planning for Tsunami: Seven Principles for Planning and Designing for Tsunami* [17], provides general guidance to local elected officials and those involved in planning, zoning, and building regulation in areas vulnerable to tsunami inundation.

As in other areas where flooding can occur, structures should be elevated above the expected tsunami inundation height. Energy-abating structures, earth berms, and vegetation can dissipate some of the energy of the incoming and receding waves, but they are not a failsafe solution. In areas where inundation is expected to be less than a meter, flood walls may protect structures from both surge and battering from debris (Figure 3-18).

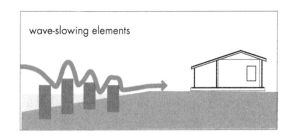

Figure 3-18
Slowing a
tsunami.

Structures in low-lying areas should be designed so that the surge passes under or through the building, by elevation of the structure or by creating "weak non-structural walls", also known as"break-away walls", perpendicular to expected waves (Figure 3-19). This would allow the waves and debris to pass through the structure. Buildings should be oriented perpendicular to wave inundation to provide the smallest profile to the wave. However, it is critical in areas that experience both tsunami and earthquakes that foundations and structures be designed to resist earthquake forces and the forces of water velocity, debris battering, and scouring and liquefaction of foundations and piles.

When program requirements such as orientation to view or site limitations necessitate building configurations that are parallel to incoming waves, attention to structural design is critical.

Figure 3-19: Avoiding a tsunami.

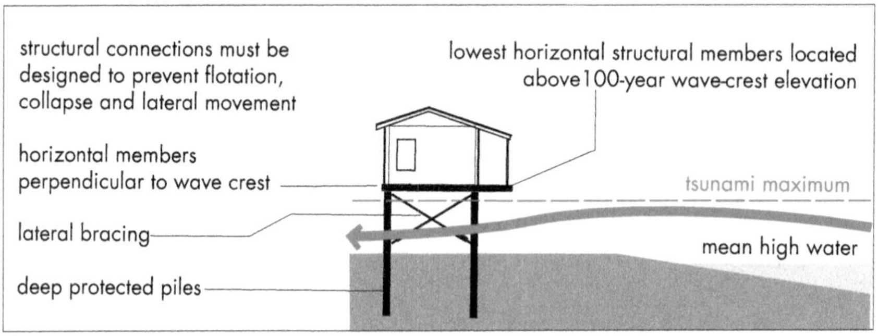

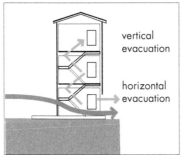

Figure 3-20

Designing evacuation options: two options are moving people to higher floors or moving people to higher ground

vertical evacuation

horizontal evacuation

Construction in coastal zones of California, Oregon, Washington, Alaska, and Hawaii must accommodate both tsunami inundation and earthquake ground motions from large near-coast events.

Site planning in areas of flood, coastal surge, and tsunami must provide for rapid evacuation of occupants to high ground, or structures must be designed for vertical evacuation to floors above forecast flood levels (Figure 3-20). This requires careful engineering because there is currently no guidance available for determining loads. For communities subject to both flood and earthquake hazards, structures intended for vertical evacuation should be designed to seismic standards higher than "life safety" so they will be available after the earthquake to accommodate tsunami evacuees.

## 3.9 CONCLUSION

The success of a project in meeting the client's expectations begins with the right team of architects and geotechnical, civil, and structural engineers. Understanding the seismic hazards in all of their direct and indirect manifestations is critical to success. Good engineering is not an excuse or a remedy for an inadequate evaluation of the site and design that does not mitigate the earthquake risk. As can be seen in the remainder of this publication, successful design is a team effort, and starts at the site.

## NOTES

1 FEMA National Flood Insurance Program. Information available at www.fema.gov/fima/

2 CGS 2002, *Guidelines for Evaluating and Mitigating Seismic hazards in Cali fornia, SpecialPublication 117*, California Geological Survey, Seismic Hazard Mapping Program, and State of California. Available at http://gmw.consrv.ca.gov/shmp/SHMPsp117.asp

3 Southern California Earthquake Center, University of Southern California, Los Angeles. Available at http://gmw.consrv.ca.gov/shmp/SHMPpgminfo.htm

4 *FEMA 386-2* is available on-line at www.fema.gov

5 Section 322, Mitigation Planning, of the Robert T. Stafford Disaster Relief and Emergency Assistance Act, enacted by Section 104 of the Disaster Mitigation Act of 2000 (Public Law 106-390)

6 §201.6(C)(2), 44 CFR PART 201, State and Local Plan Interim Criteria Under the Disaster Mitigation Act of 2000

7 Available at http://geohazards.cr.usgs.gov/eq/html/wus2002.html

8 Map available from California Seismic Safety Commission, Sacramento, CA

9 HAZUS was developed by FEMA through a cooperative agreement with the National Institute of Building Sciences. Information is available at www.fema.gov/hazus/hz_index.shtm

10 *California Public Resources Code*, Division 2, Geology, Mines and Mining, Chapter 7.5, Sections 2621-2630, Earthquake Fault Zoning. Available at www.consrv.ca.gov/cgs/rghm/ap/chp_7_5.htm#toc

11 CGS 1999, *Fault Rupture Hazard Zones in California*, Special Publication 42. Available at ftp://ftp.consrv.ca.gov/pubs/sp/SP42.PDF

12 *California Public Resources Code*, Division 2. Geology, Mines and Mining, Chapter 7.5 Earthquake Fault Zones, as amended

13 G.R. Martin and M.Lew, eOds, Southern California Earthquake Center, University of Southern California, 1999

14 USGS Open File Report 00-450, 2000

15 ASCE and Southern California Earthquake Center, June 2002

16 FEMA, *Costal Construction Manual*, 3rd Edition, FEMA 55CD. Available from FEMA at www.fema.gov

17 NOAA NTHMP 2001, *Designing for Tsunami: Seven Principles for Planning and Designing for Tsunami Hazards*. Laurence Mintier, ed. Available from http://www.pmel.noaa.gov/tsunami-hazard/links.html#multi

# EARTHQUAKE EFFECTS ON BUILDINGS 4

by Christopher Arnold

## 4.1 INTRODUCTION

This chapter explains how various aspects of earthquake ground motion affect structures and also how certain building attributes modify the ways in which the building responds to the ground motion. The interaction of these characteristics determines the overall seismic performance of the building: whether it is undamaged; suffers minor damage; becomes unusable for days, weeks, or months; or collapses with great loss of life.

Explanations of some characteristics of ground motion are followed by descriptions of several material, structural, and building attributes that, by interacting with ground motion, determine the building's seismic performance—the extent and nature of its damage.

This chapter uses the information on the nature of ground motion in Chapters 2 and 3, and applies it to structures and buildings. Chapter 5, on seismic issues in architectural design, continues the exploration of design and construction issues that, in a seismic environment, determine building performance.

## 4.2 INERTIAL FORCES AND ACCELERATION

The seismic body and surface waves create inertial forces within the building. Inertial forces are created within an object when an outside force tries to make it move if it is at rest or changes its rate or direction of motion if it is moving. Inertial force takes us back to high school physics and to Newton's Second Law of Motion, for when a building shakes it is subject to inertial forces and must obey this law just as if it were a plane, a ship, or an athlete. Newton's Second Law of Motion states that an inertial force, F, equals mass, M, multiplied by the acceleration, A. (Figure 4-1)

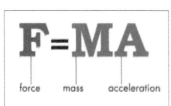

Figure 4-1:

Newton's Second Law of Motion.

Mass can be assumed as equivalent (at ground level) to the weight of the building, and so this part of the law explains why light buildings, such as wood frame houses, tend to perform better in earthquakes than large heavy ones - the forces on the building are less.

The acceleration, or the rate of change of the velocity of the waves setting the building in motion, determines the percentage of the building mass or weight that must be dealt with as a horizontal force.

Acceleration is measured in terms of the acceleration due to gravity or **g** (Figure 4-2). One g is the rate of change of velocity of a free-falling body in space. This is an additive velocity of 32 feet per second per second. Thus, at the end of the first second, the velocity is 32 feet per second; a second later it is 64 feet per second, and so on. When parachutists or bungee jumpers are in free fall, they are experiencing an acceleration of 1g. A building in an earthquake experiences a fraction of a second of g forces in one direction before they abruptly change direction.

Figure 4-2: Some typical accelerations.

The parachutists are experiencing 1g, while the roller-coaster riders reach as much as 4g. The aerobatic pilots are undergoing about 9g. The human body is very sensitive and can feel accelerations as small as 0.001g.

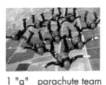
1 "g"   parachute team

4 "g"   roller coaster

0.001 "g"
human perception

9 "g" air force display team

Engineering creations (planes, ships, cars, etc.) that are designed for a dynamic or moving environment can accommodate very large accelerations. Military jet planes, for example, are designed for accelerations of up to 9g. At this acceleration, the pilot experiences 9 times the body weight pressing down on the organs and blacks out.

A commercial airliner in fairly severe turbulence may experience about 20 percent g (or 0.2g), although unbuckled passengers and attendants have been known to hit the ceiling as a result of an acceleration "drop" of over 1g. A fast moving train on a rough track may also experience up to about 0.2g.

Poorly constructed buildings begin to suffer damage at about 10 percent g (or 0.1g). In a moderate earthquake, the waves of vibration may last for a few seconds, and accelerations may be approximately 0.2g. For people on the ground or at the bottom of a building, the sensations will be very similar to those of the occupants of the plane in turbulence or passengers standing in the fast moving train: they feel unsteady and may need to grab onto something to help them remain standing. Earthquakes cause additional alarm because when the shaking starts, those experiencing it do not know whether it will quickly end or is the beginning of a damaging and dangerous quake. Short accelerations may, for a fraction of a second, exceed 1.0g. In the Northridge earthquake in 1994, a recording station in Tarzana, five miles (8 km) from the epicenter, recorded 1.92g.

## 4.3 DURATION, VELOCITY, AND DISPLACEMENT

Because of the inertial force formula, acceleration is a key factor in determining the forces on a building, but a more significant measure is that of acceleration combined with duration, which takes into account the impact of earthquake forces over time. In general, a number of cycles of moderate acceleration, sustained over time, can be much more difficult for a building to withstand than a single much larger peak. Continued shaking weakens a building structure and reduces its resistance to earthquake damage.

A useful measure of strong-motion duration is termed the **bracketed duration**. This is the shaking duration above a certain threshold acceleration value, commonly taken as 0.05g, and is defined as the time between the first and last peaks of motion that exceeds this threshold value. In the San Fernando earthquake of 1971, the bracketed duration was only about 6 seconds. In both the Loma Prieta and the Northridge earthquakes, the strong motion lasted a little over ten seconds, yet caused much destruction. In the 1906 San Francisco earthquake, the severe shaking lasted 45 seconds, while in Alaska, in 1964, the severe motion lasted for over three minutes.

Two other measures of wave motion are directly related to acceleration and can be mathematically derived from it. Velocity, which is measured in inches or centimeters per second, refers to the rate of motion of the seismic waves as they travel through the earth. This is very fast. Typically, the **P wave** travels at between 3 km/sec and 8 km/sec or 7,000 to 18,000 mph. The **S wave** is slower, traveling at between 2 km/sec and 5 km/sec, or 4,500 mph to 11,000 mph.

Displacement refers to the distance that points on the ground are moved from their initial locations by the seismic waves. These distances, except immediately adjacent to or over the fault rupture, are quite small and are measured in inches or centimeters. For example, in the Northridge earthquake, a parking structure at Burbank, about 18 miles (29 km) from the epicenter recorded displacements at the roof of 1.6 inches (4.0 cm) at an acceleration of 0.47g. In the same earthquake, the Olive View hospital in Sylmar, about 7.5 miles (12 km) from the epicenter, recorded a roof displacement of 13.5 inches (34 cm) at an acceleration of 1.50g.

The velocity of motion on the ground caused by seismic waves is quite slow—huge quantities of earth and rock are being moved. The velocity varies from about 2 cm/sec in a small earthquake to about 60 cm/sec in a major shake. Thus, typical building motion is slow and the distances are small, but thousands of tons of steel and concrete are wrenched in all directions several times a second.

In earthquakes, the values of ground acceleration, velocity, and displacement vary a great deal in relation to the frequency of the wave motion. High–frequency waves (higher than 10 hertz) tend to have high amplitudes of acceleration but small amplitudes of displacement, compared to low-frequency waves, which have small accelerations and relatively large velocities and displacements.

## 4.4 GROUND AMPLIFICATION

Earthquake shaking is initiated by a fault slippage in the underlying rock. As the shaking propagates to the surface, it may be amplified, depending on the intensity of shaking, the nature of the rock and, above all, the surface soil type and depth.

A layer of soft soil, measuring from a few feet to a hundred feet or so, may result in an amplification factor of from 1.5 to 6 over the rock shaking. This amplification is most pronounced at longer periods, and may not be so significant at short periods. (Periods are defined and explained in the next section, 4.5.1.) The amplification also tends to decrease as the level of shaking increases.

As a result, earthquake damage tends to be more severe in areas of soft ground. This characteristic became very clear when the 1906 San Francisco earthquake was studied, and maps were drawn that showed building damage in relation to the ground conditions. Inspection of records from soft clay sites during the 1989 Loma Prieta earthquake indicated a maximum amplification of long-period shaking of three to six times. Extensive damage was caused to buildings in San Francisco's Marina district, which was largely built on filled ground, some of it rubble deposited after the 1906 earthquake.

Because of the possibility of considerable shaking amplification related to the nature of the ground, seismic codes have some very specific requirements that relate to the characteristics of the site. These require the structure to be designed for higher force levels if it is located on poor soil. Specially designed foundations may also be necessary.

## 4.5 PERIOD AND RESONANCE

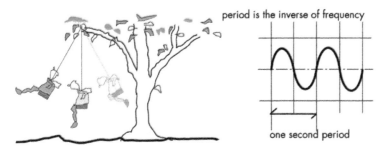

Figure 4-3 The Fundamental Period.

### 4.5.1 Natural Periods

Another very important characteristic of earthquake waves is their **period** or frequency; that is, whether the waves are quick and abrupt or slow and rolling. This phenomenon is particularly important for determining building seismic forces.

All objects have a natural or fundamental period; this is the rate at which they will move back and forth if they are given a horizontal push (Figure 4-3). In fact, without pulling and pushing it back and forth, it is not possible to make an object vibrate at anything other than its natural period.

When a child in a swing is started with a push, to be effective this shove must be as close as possible to the natural period of the swing. If correctly gauged, a very small push will set the swing going nicely. Similarly, when earthquake motion starts a building vibrating, it will tend to sway back and forth at its natural period.

Period is the time in seconds (or fractions of a second) that is needed to complete one cycle of a seismic wave. Frequency is the inverse of this—the number of cycles that will occur in a second—and is measured in "Hertz". One Hertz is one cycle per second.

Natural periods vary from about 0.05 seconds for a piece of equipment, such as a filing cabinet, to about 0.1 seconds for a one-story building. Period is the inverse of frequency, so the cabinet will vibrate at 1 divided by 0.05 = 20 cycles a second or 20 Hertz.

A four-story building will sway at about a 0.5 second period, and taller buildings between about 10 and 20 stories will swing at periods of about

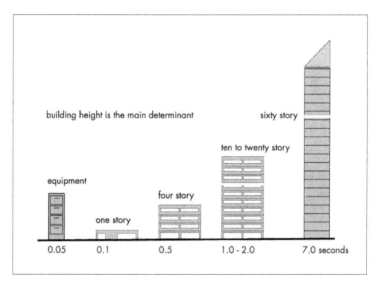

Figure 4-4: Comparative building periods, determined by height. These values are approximations: the structural system, materials, and geometric proportions will also affect the period.

1 to 2 seconds. A large suspension bridge may have a period of around 6 seconds. A rule of thumb is that the building period equals the number of stories divided by 10; therefore, period is primarily a function of building height. The 60-story Citicorp office building in New York has a measured period of 7 seconds; give it a push, and it will sway slowly back and forth completing a cycle every 7 seconds. Other factors, such as the building's structural system, its construction materials, its contents, and its geometric proportions, also affect the period, but height is the most important consideration (Figure 4-4).

The building's period may also be changed by earthquake damage. When a reinforced concrete structure experiences severe ground shaking, it begins to crack: this has the effect of increasing the structure's period of vibration: the structure is "softening". This may result in the structure's period approaching that of the ground and experiencing resonance, which may prove fatal to an already weakened structure. The opposite effect may also occur: a steel structure may stiffen with repeated cycles of movement until the steel yields and deforms.

## 4.5.2 Ground Motion, Building Resonance, and Response Spectrum

When a vibrating or swinging object is given further pushes that are also at its natural period, its vibrations increase dramatically in response to even rather small pushes and, in fact, its accelerations may increase as much as four or five times. This phenomenon is called resonance.

The ground obeys the same physical law and also vibrates at its natural period, if set in motion by an earthquake. The natural period of ground varies from about 0.4 seconds to 2 seconds, depending on the nature of the ground. Hard ground or rock will experience short period vibration. Very soft ground may have a period of up to 2 seconds but, unlike a structure, it cannot sustain longer period motions except under certain unusual conditions. Since this range is well within the range of common building periods, it is quite possible that the pushes that earthquake ground motion imparts to the building will be at the natural period of the building. This may create resonance, causing the structure to encounter accelerations of perhaps 1g when the ground is only vibrating with accelerations of 0.2g. Because of this, buildings suffer the greatest damage from ground motion at a frequency close or equal to their own natural frequency.

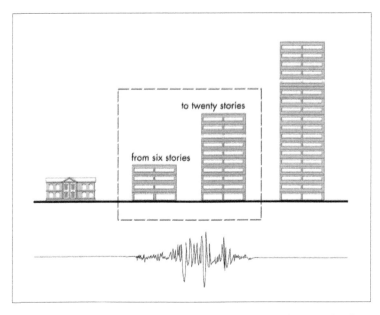

Figure 4-5: The vulnerable group: Mexico City, 1985. The periods of buildings in the 6 to 20 story range resonated with the frequency content of the earthquake.

The terrible destruction in Mexico City in the earthquake of 1985 was primarily the result of response amplification caused by coincidence of building and ground motion periods (Figure 4-5). Mexico City was some 250 miles from the earthquake focus, and the earthquake caused the soft ground in margins of the old lake bed under the downtown buildings to vibrate for over 90 seconds at its long natural period of around 2 seconds. This caused buildings that were between about 6 and 20 stories in height to resonate at a similar period, greatly increasing the accelerations within them. Taller buildings suffered little damage. This amplification in building vibration is very undesirable. The possibility of it happening can be reduced by trying to ensure that the building period will not coincide with that of the ground. Thus, on soft (long-period) ground, it would be best to design a short, stiff (short-period) building.

Taller buildings also will undergo several modes of vibration so that the building will wiggle back and forth like a snake (Figure 4-6).

However, later modes of vibration are generally less critical than the natural period, although they may be significant in a high-rise building. For low-rise buildings, the natural period (which, for common structures, will always be relatively short) is the most significant. Note, however, that the low-period, low- to mid-rise building is more likely to experience resonance from the more common short-period ground motion.

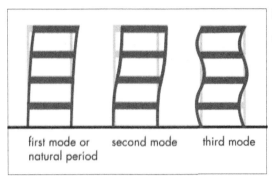

Figure 4-6

Modes of vibration.

first mode or natural period     second mode     third mode

### 4.5.3 Site Response Spectrum

From the above, it can be seen that buildings with different periods (or frequency responses) will respond in widely differing ways to the same earthquake ground motion. Conversely, any building will act differently during different earthquakes, so for design purposes it is necessary to represent the building's range of responses to ground motion of different frequency content. Such a representation is termed a site **response spectrum**. A site response spectrum is a graph that plots the maximum response values of acceleration, velocity, and displacement against period (and frequency). Response spectra are very important tools in earthquake engineering.

Figure 4-7 shows a simplified version of a response spectrum. These spectra show, on the vertical ordinate, the accelerations, velocities and displacements that may be expected at varying periods (the horizontal ordinate). Thus, the response spectrum illustrated shows a maximum acceleration response at a period of about 0.3 seconds—the fundamental period of a midrise building. This shows how building response varies with building period: as the period lengthens, accelerations decrease and displacement increases. On the other hand, one- or two-story buildings with short periods undergo higher accelerations but smaller displacements.

In general, a more flexible longer period design may be expected to experience proportionately lesser accelerations than a stiffer building. A glance at a response spectrum will show why this is so: as the period of the building lengthens (moving towards the right of the horizontal axis of the spectrum), the accelerations reduce. Currently our codes recognize the beneficial aspect of flexibility (long period) by permitting lower design coefficients. However, there is an exchange, in that the lower accelerations in the more flexible design come at the expense of more motion. This increased motion may be such that the building may suffer considerable damage to its nonstructural components, such as ceilings and partitions, in even a modest earthquake.

Figure 4-7

Simplified response spectra, for acceleration, velocity and displacement.

SOURCE MCEER INFORMATION SERVICE

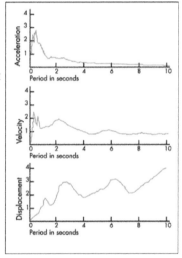

Seismic codes provide a very simple standardized response spectrum that is suitable for small buildings (Figure 4-8). The code also provides a procedure for the engineer to construct a more accurate response spectrum for the building, based on various assumptions. For larger buildings in which a geotechnical consultant provides information on the site characteristics and an estimate of ground motions, detailed response spectra will also be provided to assist the engineer in the calculation of forces that the building will encounter.

The response spectrum enables the engineer to identify the resonant frequencies at which the building will undergo peak accelerations. Based

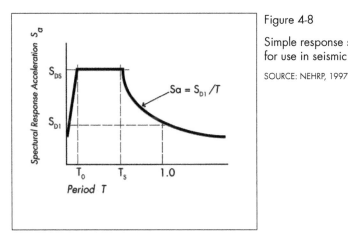

Figure 4-8

Simple response spectrum, for use in seismic codes.

SOURCE: NEHRP, 1997

on this knowledge, the building design might be adjusted to ensure that the building period does not coincide with the site period of maximum response. For the site characteristics shown, with a maximum response at about 0.3 seconds, it would be appropriate to design a building with a longer period of 1 second or more. Of course, it is not always possible to do this, but the response spectrum shows clearly what the possible accelerations at different periods are likely to be, and the forces can be estimated more accurately. Information gained from a response spectrum is of most value in the design of large and high structures.

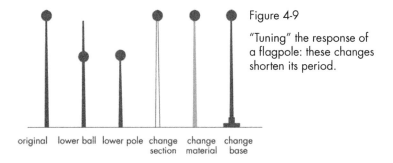

Figure 4-9

"Tuning" the response of a flagpole: these changes shorten its period.

How does one "tune" a building, or change its period, if it is necessary to do so? One could change the natural period of a simple structure such as a flag pole by any or all combinations of the methods shown in Figure 4-9:

○ Changing the position of the weight to a lower height

○ Changing the height of the pole

○ Changing the sectional area or shape of the pole

○ Changing the material

○ Altering the fixity of the base anchorage

There are analogous possibilities for buildings, though the building is much more complex than the simple monolithic flagpole:

○ Tune the building by ensuring that the structural characteristics of a design are compatible with those of the site as the preliminary design is developed.

○ Incorporate devices in the structure that dissipate energy and change the response characteristics. These devices are discussed in Chapter 7.

○ After the Mexico City earthquake of 1985, a number of damaged buildings had their upper floors removed to lower their period and reduce their mass, thus reducing the likelihood and consequences of resonance.

## 4.6 DAMPING

If a structure is made to vibrate, the amplitude of the vibration will decay over time and eventually cease. Damping is a measure of this decay in amplitude, and it is due to internal friction and absorbed energy. The nature of the structure and its connections affects the damping; a heavy concrete structure will provide more damping than a light steel frame. Architectural features such as partitions and exterior façade construction contribute to the damping.

Damping is measured by reference to a theoretical damping level termed **critical damping**. This is the least amount of damping that will allow the structure to return to its original position without any continued vibration. For most structures, the amount of damping in the system will vary from between 3 percent and 10 percent of critical. The higher values would apply to older buildings (such as offices and government buildings) that employed a structure of steel columns and beams encased in concrete together with some structural walls, which also had many heavy fixed partitions (often concrete block or hollow tile), and would have high damping values. The lower values would apply to a modern office building with a steel-moment frame, a light metal and glass exterior envelope, and open office layouts with a minimum of fixed partitions.

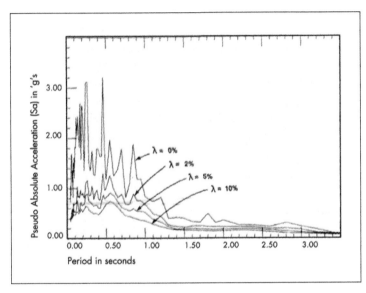

Figure 4-10

Response spectra for a number of damping values.

SOURCE: STRATTA, 1987

The main significance of damping is that accelerations created by ground motion increase rapidly as the damping value decreases. The response spectra shown in Figure 4-10 show that the peak acceleration is about 3.2g for a damping value of 0 %, 0.8g for a damping value of 2 % and a value of about 0.65g for a value of 10 %.

Tables are available that indicate recommended damping values, and the damping characteristics of a structure can be fairly easily estimated. Response spectra generally show acceleration values for 0, 2, 5, and 10 % damping. A damping value of zero might be used in the design of an simple vibrator, such as a flag pole or a water tank supported on a single cantilever column. For typical structures, engineers generally use a value of 5 % critical.

Damping used to be regarded as a fixed attribute of buildings, but in recent years a number of devices have been produced that enable the engineer to increase the damping and reduce the building response. This greatly increases the designer's ability to provide a "tuned" response to the ground motion.

## 4.7 DYNAMIC AMPLIFICATION

It was early observed and calculated that for most structures, the structural movement is greater than that of the ground motion. The increase of the structural movement over that of the ground motion is commonly referred to as **dynamic amplification**. This amplification is caused when energy is reflected from the P and S waves when they reach the earth's surface, which is consequently affected almost simultaneously by upward and downward moving waves (See Chapter 2, Section 2.3.2). The extent of amplification also varies depending on the dynamic properties of the structure and the characteristics of the initial earthquake ground motion encountered. The important engineering attributes of the structure are:

○ The period of vibration of the structure

○ The damping properties of the structure

For typical earthquake motions and for structures having the common damping value of 5 percent damping and a period range of 0.5 seconds to 3.3 seconds, the dynamic amplification factor would be about 2.5, which is a significant increase. For higher damping values, the amplification factor is reduced.

## 4.8 HIGHER FORCES AND UNCALCULATED RESISTANCE

Even if a building is well damped and is unlikely to resonate, it may be subjected to forces that are much higher than the computed forces for which it was designed. For those familiar with the assumptions of design for vertical loads and the large factors of safety that are added into the calculations, this may seem surprising. Why is this so?

The answer is that to design a building for the very rare maximum conceivable earthquake forces, and then to add a factor of safety of two or three times as is done for vertical loads, would result in a very expensive structure whose functional use would be greatly compromised by massive structural members and structural walls with very limited openings: the ordinary building would resemble a nuclear power plant or a military bunker.

Experience has shown, however, that many buildings have encountered forces far higher than they were designed to resist and yet have survived, sometimes with little damage. This phenomenon can be explained by the fact that the analysis of forces is not precise and deliberately errs on the conservative side so that the building strength is, in reality, greater than the design strength. In addition, the building often gains additional strength from components, such as partitions, that are not considered in a structural analysis. Some structural members may be sized for adequate stiffness rather than for strength, and so have considerable reserve strength. Materials often are stronger in reality than the engineer assumes in his calculations. Finally, seismically engineered structures have an additional characteristic that acts to provide safety in the event of encountering forces well beyond the design threshold: this is the important property of **ductility**. Taken together, these characteristics, though not all explicit, provide a considerable safety factor or uncalculated additional resistance.

## 4.9 DUCTILITY

The gap between design capacity (the theoretical ability of a building to withstand calculated forces) and possible actual forces is, finally, largely dealt with by relying on the material property of ductility. This is the property of certain materials (steel in particular) to fail only after considerable inelastic deformation has taken place, meaning that the material does not return to its original shape after distortion. This deformation, or distortion, dissipates the energy of the earthquake.

This is why it is much more difficult to break a metal spoon by bending it than one made of plastic. The metal object will remain intact, though distorted, after successive bending to and fro while the plastic spoon will snap suddenly after a few bends. The metal is far more ductile than the plastic (Figure 4-11).

The deformation of the metal (even in the spoon) absorbs energy and defers absolute failure of the structure. The material bends but does not break and so continues to resist forces and support loads, although with diminished effectiveness. The effect of earthquake motion on a building

Figure 4-11: Ductility

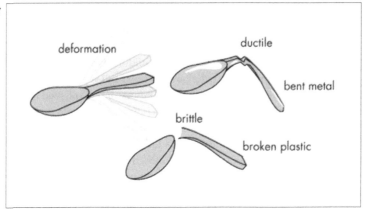

is rather like that of bending a spoon rapidly back and forth: the heavy structure is pushed back and forth in a similar way several times a second (depending on its period of vibration).

Brittle materials, such as unreinforced masonry or inadequately reinforced concrete, fail suddenly, with a minimum of prior distortion. The steel bars embedded in reinforced concrete can give this material considerable ductility, but heavier and more closely spaced reinforcing bars and special detailing of their placement are necessary.

Ductility and reserve capacity are closely related: past the elastic limit (the point at which forces cause permanent deformation), ductile materials can take further loading before complete failure. In addition, the member proportions, end conditions, and connection details will also affect ductility. Reserve capacity is the ability of a complete structure to resist overload, and is dependent on the ductility of its individual members. The only reason for not requiring ductility is to provide so much resistance that members would never exceed elastic limits.

Thus, buildings are designed in such a way that in the rare case when they are subjected to forces higher than those required by a code, the materials and connections will distort but not break. In so doing, they will safely absorb the energy of the earthquake vibrations, and the building, although distorted and possibly unusable, is at least still standing.

## 4.10 STRENGTH, STIFFNESS, FORCE DISTRIBUTION, AND STRESS CONCENTRATION

### 4.10.1 Strength and Stiffness

Strength and stiffness are two of the most important characteristics of any structure. Although these two concepts are present in non-seismic structural design and analysis, the distinction between strength and stiffness is perhaps most critical, and its study most highly developed, in structural engineering for lateral forces.

Sufficient strength is necessary to ensure that a structure can support imposed loads without exceeding certain stress values. Stress refers to the internal forces within a material or member that are created as the structural member resists the applied load. Stress is expressed in force per unit area (for example, pounds per square inch).

Stiffness is measured by deflection, the extent to which a structural member, such as a floor, roof, or wall structure, bends when loaded. Deflection is generally expressed as a fraction of length of the member or assembly. For gravity loads, this is usually the only aspect of stiffness that is of concern. When floor joists are designed for a house, for example, it is often deflection rather than strength that dictates the size of the joists—that is, the depth of the joists is determined by how much they will bend under load rather than by whether they can safely support the floor loads. Typically, an unacceptable amount of bending (in the form of an uncomfortable "springy" feeling to occupants) will occur well before the joists are stressed to the point at which they may break because of overload.

To ensure sufficient strength and stiffness, codes such as the *International Building Code* (IBC) provide stress and deflection limits that are not to be exceeded for commonly used materials and assemblies. For example, interior partitions "shall be designed to resist all loads to which they are subjected, but not less than a force of 5 pounds per square foot applied perpendicular to the walls." In addition, "the deflection of such walls under a load of 5 pounds per square foot shall not exceed 1/240 of the span for walls with brittle finishes and 1/120 of the span for walls with flexible finishes."

Most designers are familiar with deflection in this sense, and have an intuitive feel for this quality. In seismic design, deflection of vertical structural members, such as columns and walls, is termed **drift**. Analogous to the deflection of horizontal members, limitations on drift may impose more severe requirements on members than the strength requirements. Story drift is expressed as the difference of the deflections at the top and bottom of the story under consideration: this is also often expressed as a ratio between the deflection and the story, or floor-to-floor height (Figure 4-12). Drift limits serve to prevent possible damage to interior or exterior walls that are attached to the structure and which might be cracked or distorted if the structure deflects too much laterally, creating racking forces in the member. Thus the IBC requires that drift be limited in typical buildings to between 0.02 and 0.01 times the building height, depending on the occupancy of the building. For a building that is 30 feet high, drift would be limited to between 3.6 inches and 7.2 inches depending on the building type.

Figure 4-12
Story drift ratio.

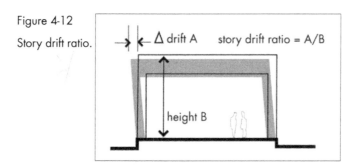

When the earthquake-induced drift is excessive, vertical members may become permanently deformed; excessive deformation can lead to structural and nonstructural damage and, ultimately, collapse.

Thus strength and stiffness are two important characteristics of any structural member. Two structural beams may have approximately equal material strengths and be of similar shape but will vary in their stiffness and strength, depending on how they are oriented relative to the load. This concept can be easily understood by visualizing the flexibility of a narrow, deep beam placed where it has to support a load: the extent of deflection will depend on whether the load is placed on the beams flat surface or on its edge (Figure 4-13).

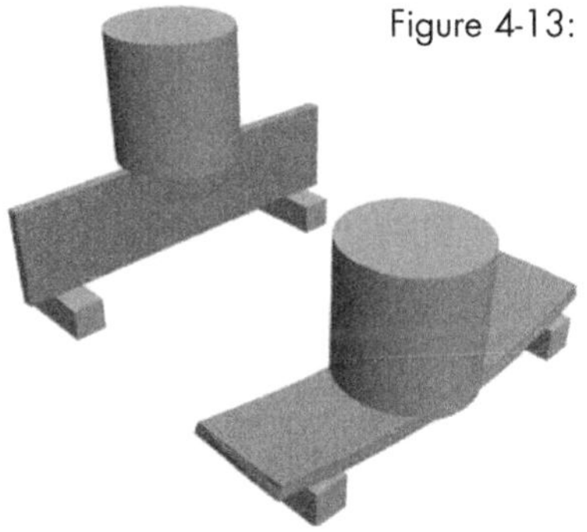

Figure 4-13: Strength and stiffness.

members are approximately equal in strength but their stiffnesses are different.

lateral forces are distributed in proportion to the stiffness of the resisting members.

## 4.10.2 Force Distribution and Stress Concentration

In seismic design, there is another very important characteristic of stiffness, besides that of deflection. The simple solution of determining the overall lateral force on the building by multiplying the building weight by its acceleration has already been discussed. But the engineer needs to know how this force is allocated to the various resisting structural elements that must be designed: each shares some proportion of this overall force. The answer is that the force is distributed in proportion to the **relative stiffness of the resisting members**. In other terms, the applied forces are "attracted to" and concentrated at the stiffer elements of the building. Thus the engineer must calculate the stiffness of the resisting elements to ascertain the forces that they must accommodate.

The relative rigidities of members are a major concern in seismic analysis. As soon as a rigid horizontal element or diaphragm, such as a concrete slab, is tied to vertical resisting elements, it will force those elements to deflect the same amount. If two elements (two frames, walls, braces, or any combination) are forced to deflect the same amount, and if one is stiffer, that one will take more of the load. Only if the stiffnesses are identical can it be assumed that they share the load equally. Since concrete slab floors or roofs will generally fit into the "rigid diaphragm" classification, and since it is unusual for all walls, frames, or braced frames to be identical, the evaluation of relative rigidities is a necessary part of most seismic analysis problems in order to determine the relative distribution of the total horizontal force to the various resisting elements.

The reason why forces are related to the stiffness of the resisting elements can be understood by visualizing a heavy block supported away from a wall by two short beams. Clearly, the thick, stiff beam will carry much more load than the slender one, and the same is true if they are turned 90 degrees to simulate the lateral force situation (Figure 4-14).

Figure 4-14

Force distribution and stiffness.

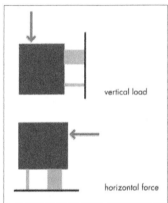

vertical load

horizontal force

An important aspect of this concept in relation to column lateral stiffness is illustrated in Figure 4-15. In this figure the columns have the same cross-section, but the short column is half the length of the long one. Mathematically, the stiffness of a column varies approximately as the cube of its length. Therefore, the short column will be eight times stiffer ($2^3$) instead of twice as stiff and will be subject to eight times the horizontal load of the long column. Stress is concentrated in the short column, while the long column is subject to nominal forces.

In a building with members of varying stiffness, an undue proportion of the overall forces may be concentrated at a few points of the building, such as a particular set of beams, columns, or walls, as shown at the top of Figure 4-15. These few members may fail and, by a chain reaction, bring down the whole building. People who are in the building demolition business know that if they weaken a few key columns or connections in a building, they can bring it down. An earthquake also tends to "find" these weak links.

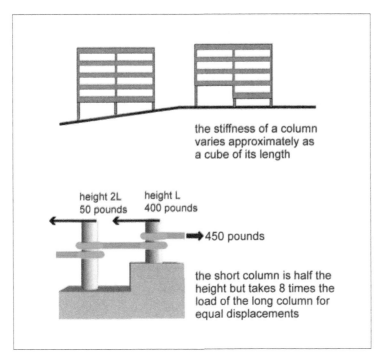

the stiffness of a column
varies approximately as
a cube of its length

height 2L          height L
50 pounds        400 pounds

450 pounds

the short column is half the
height but takes 8 times the
load of the long column for
equal displacements

Figure 4-15: The short column problem.

This condition has serious implications for buildings with column or shear walls of different length. In designing a structure, the engineer tries to equalize the stiffness of the resisting elements so that no one member or small group of members takes a disproportionate amount of the load. If this cannot be done - if their size and stiffness vary for architectural reasons, for example – then the designer must make sure that stiffer members are appropriately designed to carry their proportion of the load.

A special case of this problem is that sometimes a short-column condition is created inadvertently after the building is occupied. For example, the space between columns may be filled in by a rigid wall, leaving a short space for a clerestory window. Such a simple act of remodeling may not seem to require engineering analysis, and a contractor may be hired to do the work: often such work is not subject to building department reviews and inspection. Serious damage has occurred to buildings in earthquakes because of this oversight (Figure 4-16).

Figure 4-16

Creation of inadvertent short columns.

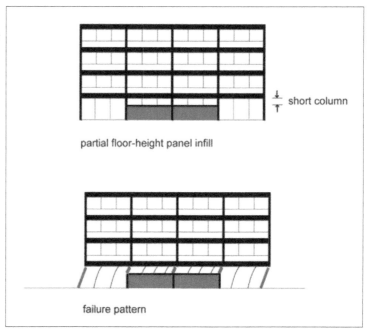

partial floor-height panel infill

short column

failure pattern

## 4.11 TORSIONAL FORCES

The center of mass, or center of gravity, of an object is the point at which it could be exactly balanced without any rotation resulting. If the mass (or weight) of a building is uniformly distributed (in plan), the result is that the plan's geometric center will coincide with the center of mass. In a building, the main lateral force is contributed by the weight of the floors, walls, and roof, and this force is exerted through the center of mass, usually the geometric center of the floor (in plan). If the mass within a floor is uniformly distributed, then the resultant force of the horizontal acceleration of all its particles is exerted through the floor's geometric center. If the resultant force of the resistance (provided by shear walls, moment frames, or braced frames) pushes back through this point, dynamic balance is maintained.

Torsional forces are created in a building by a lack of balance between the location of the resisting elements and the arrangement of the building mass. Engineers refer to this as **eccentricity between the center of mass and the center of resistance,** which makes a building subjected to ground motion rotate around its center of resistance, creating torsion

- a twisting action in plan, which results in undesirable and possibly dangerous concentrations of stress (Figure 4-17).

In a building in which the mass is approximately evenly distributed in plan (typical of a symmetrical building with uniform floor, wall and column masses) the ideal arrangement is that the earthquake resistant elements should be symmetrically placed, in all directions, so that no matter in which direction the floors are pushed, the structure pushes back with a balanced stiffness that prevents rotation from trying to occur. This is the reason why it is recommended that buildings in areas of seismic risk be designed to be as symmetrical as possible. In practice, some degree of torsion is always present, and the building code makes provision for this.

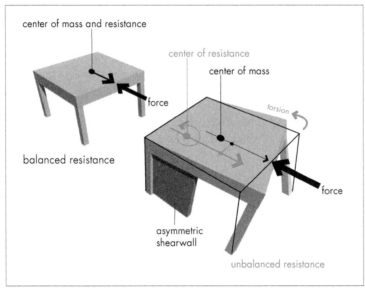

Figure 4-17
Torsional forces.

## 4.12 NONSTRUCTURAL COMPONENTS

For many decades, seismic building codes focused exclusively on the structure of the building—that is, the system of columns, beams, walls, and diaphragms that provides resistance against earthquake forces. Although this focus remains dominant for obvious reasons, experience

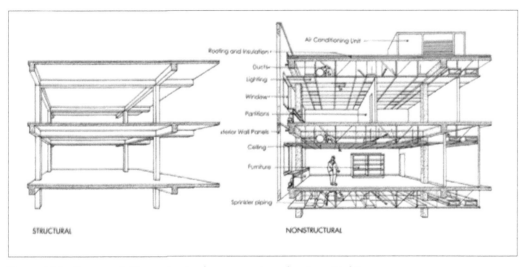

Figure 4-18: Structure (left), nonstructural components and systems (right).

in more recent earthquakes has shown that damage to nonstructural components is also of great concern. In most modern buildings, the nonstructural components account for 60 to 80 percent of the value of the building (Figure 4-18). Most nonstructural components are fragile (compared to the building structure), easily damaged, and costly to repair or replace (Figure 4-19).

Figure 4-19

Typical nonstructural damage in school. Northridge earthquake, 1994.

The distinction between structural and nonstructural components and systems is, in many instances, artificial. The engineer labels as nonstructural all those components that are not designed as part of the seismic lateral force-resisting system. Nature, however, makes no such distinction, and tests the whole building. Many nonstructural components may be called upon to resist forces even though not designed to do so.

The nonstructural components or systems may modify the structural response in ways detrimental to the safety of the building. Examples are the placing of heavy nonstructural partitions in locations that result in severe torsion and stress

EARTHQUAKE EFFECTS ON BUILDINGS

concentration, or the placement of nonstructural partitions between columns in such a way as to produce a short column condition, as described in Section 4.10 and illustrated in Figure 4-16. This can lead to column failure, distortion, and further nonstructural damage. Failure of the fire protection system, because of damage to the sprinkler system, may leave the building vulnerable to post-earthquake fires caused by electrical or gas system damage.

## 4.13 CONSTRUCTION QUALITY

One other characteristic that applies to any structure may be obvious, but must be emphasized: the entire structural system must be correctly constructed if it is to perform well. Lateral forces are especially demanding because they actively attempt to tear the building apart, whereas vertical loads (with the exception of unusual live loads such as automobiles or large masses of people) sit still and quiet within the materials of the building.

The materials of the seismically resistant structure must have the necessary basic strength and expected properties, but most importantly, all the structural components must be securely connected together so that as they push and pull against one another during the earthquake, the connections are strong enough to transfer the earthquake forces and thereby maintain the integrity of the structure.This means that detailed design and construction of connections are particularly important.

The correct installation of reinforcing steel and anchors in concrete structures; the correct design, fabrication and installation of connection members in steel structures; and correct nailing, edge clearances and installation of hold-downs in wood framing are all critical. For nonstructural components, critical issues are the maintenance of correct clearances between precast concrete cladding panels, at seismic separation joints, and between glazing and window framing, and the correct design and installation of bracing of heavy acceleration-sensitive components such as tanks, chillers, heavy piping, electrical transformers and switch gear.

Quality control procedures must be enforced at all phases of design and construction, including material testing and on-site inspection by qualified personnel. The earthquake is the ultimate testing laboratory of the construction quality of the building.

## 4.14 CONCLUSION

This chapter has discussed a number of key characteristics of earthquake ground shaking that affect the seismic performance of buildings. In addition, a number of building characteristics have been reviewed that, together with those of the ground, determine the building's seismic performance: how much damage the building will suffer. These characteristics are common to all buildings, both new and existing, and all locations.

The building response to earthquake shaking occurs over the time of a few seconds. During this time, the several types of seismic waves are combining to shake the building in ways that are different in detail for each earthquake. In addition, as the result of variations in fault slippage, differing rock through which the waves pass, and the different geological nature of each site, the resultant shaking at each site is different. The characteristics of each building are different, whether in size, configuration, material, structural system, method of analysis, age, or quality of construction: each of these characteristics affects the building response.

In spite of the complexity of the interactions between the building and the ground during the few seconds of shaking there is broad understanding of how different building types will perform under different shaking conditions. This understanding comes from extensive observation of buildings in earthquakes all over the world, together with analytical and experimental research at many universities and research centers.

Understanding the ground and building characteristics discussed in this chapter is essential to give designers a "feel" for how their building will react to shaking, which is necessary to guide the conceptual design of their building. The next chapter continues this direction by focusing on certain architectural characteristics that influence seismic performance - either in a positive or negative way.

## 4.15 REFERENCES

Building Seismic Safety Council, *NEHRP Recommended Provisions for Seismic Regulations for New Buildings and Other Structures*, Part 2 Commentary, FEMA 450, 2003, Washington, DC.

International Code Council, *International Building Code 2003* (IBC), Birmingham, AL.

Stratta, James, *Manual of Seismic Design*,1987, Prentice-Hall, Englewood Cliffs, NJ

## 4.16 TO FIND OUT MORE

MCEER, Center for Earthquake Engineering Research.
mceer@buffalo.edu

Bolt, B. A., *Earthquakes* (4th edition) W. H. Freeman and Company New York N. Y. 2002.

Gere J. M. and Shah, H.C.,1984 *Terra non Firma*, Stanford Alumni Association, Stanford CA.

Levy M. and Salvadori M., *Why Buildings Stand Up*, W. W. Norton and Company, New York, NY. 1990.

Levy M. and Salvadori M., *Why the Earth Quakes*, W. W. Norton and Company, New York, NY. 1997.

Levy M. and Salvadori M., *Why Buildings Fall Down*, W. W. Norton and Company, New York, NY 2002.

by Christopher Arnold

## 5.1 INTRODUCTION

This chapter uses the information in the preceding chapter to explain how architectural design decisions influence a building's likelihood to suffer damage when subjected to earthquake ground motion. The critical design decisions are those that create the building configuration, defined as the building's size and three dimensional shape, and those that introduce detailed complexities into the structure, in ways that will be discussed later.

In sections 5.2 to 5.5, the effects of architectural design decisions on seismic performance are explained by showing a common structural/architectural configuration that has been designed for near optimum seismic performance and explaining its particular characteristics that are seismically desirable. In Section 5.3, the two main conditions created by configuration irregularity are explained. In Section 5.4, a number of deviations from these characteristics (predominantly architectural in origin) are identified as problematical from a seismic viewpoint. Four of these deviations are then discussed in more detail in Section 5. 5 both from an engineering and architectural viewpoint, and conceptual solutions are provided for reducing or eliminating the detrimental effects. Section 5.6 identifies a few other detailed configuration issues that may present problems.

Section 5.7 shows how seismic configuration problems originated in the universal adoption of the "International Style" in the twentieth century, while Section 5.8 gives some guidelines on how to avoid architectural/structural problems. Finally, Section 5.9 looks to the future in assessing today's architectural trends, their influence on seismic engineering, and the possibility that seismic needs might result in a new "seismic architecture".

## 5.2 THE BASIC SEISMIC STRUCTURAL SYSTEMS

A building's structural system is directly related to its architectural configuration, which largely determines the size and location of structural elements such as walls, columns, horizontal beams, floors, and roof structure. Here, the term **structural/architectural configuration** is used to represent this relationship.

### 5.2.1 The Vertical Lateral Resistance Systems

Seismic designers have the choice of three basic alternative types of vertical lateral force–resisting systems, and as discussed later, the system must be selected at the outset of the architectural design process. Here, the intent is to demonstrate an optimum architectural/structural configuration for each of the three basic systems. The three alternatives are illustrated in Figure 5-1.

These basic systems have a number of variations, mainly related to the structural materials used and the ways in which the members are connected. Many of these are shown in Chapter 7: Figures 7-2, 7-3, 7-11A and 7-11b show their comparative seismic performance characteristics.

○ Shear walls

Shear walls are designed to receive lateral forces from diaphragms and transmit them to the ground. The forces in these walls are predominantly shear forces in which the material fibers within the wall try to slide past one another. To be effective, shear walls must run from the top of the building to the foundation with no offsets and a minimum of openings.

○ Braced frames

Braced frames act in the same way as shear walls; however, they generally provide less resistance but better ductility depending on their detailed design. They provide more architectural design freedom than shear walls.

There are two general types of braced frame: conventional concentric and eccentric. In the concentric frame, the center lines of the bracing members meet the horizontal beam at a single point.

In the eccentric braced frame, the braces are deliberately designed to meet the beam some distance apart from one another: the short piece of beam between the ends of the braces is called a link beam. The purpose of the link beam is to provide ductility to the system: under heavy seismic forces, the link beam will distort and dissipate the energy of the earthquake in a controlled way, thus protecting the remainder of the structure (Figure 5-2).

SEISMIC ISSUES IN ARCHITECTURAL DESIGN

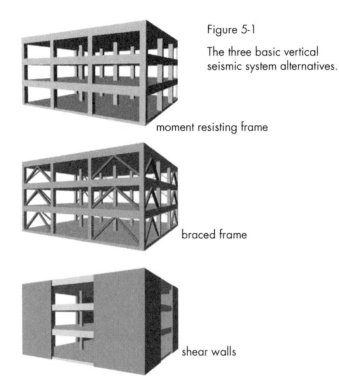

Figure 5-1

The three basic vertical seismic system alternatives.

moment resisting frame

braced frame

shear walls

○ Moment-resistant frames

A moment resistant frame is the engineering term for a frame structure with no diagonal bracing in which the lateral forces are resisted primarily by bending in the beams and columns mobilized by strong joints between columns and beams. Moment-resistant frames provide the most architectural design freedom.

These systems are, to some extent, alternatives, although designers sometimes mix systems, using one type in one direction and another type in the other. This must be done with care, however, mainly because the different systems are of varying stiffness (shear-wall systems are much stiffer than moment-resisting frame systems, and braced systems fall in between), and it is difficult to obtain balanced resistance when they are mixed. However, for high-performance structures,) there is now increasing use of dual systems, as described in section 7.7.6. Examples of effective mixed systems are the use of a shear-wall core together with a perimeter moment-resistant frame or a perimeter steel-moment frame

Figure 5-2

Types of braced frames.

concentric brace

eccentric brace
with link beams

damage limited to link beam

with interior eccentric-braced frames. Another variation is the use of shear walls combined with a moment-resistant frame in which the frames are designed to act as a fail-safe back-up in case of shear-wall failure.

The framing system must be chosen at an early stage in the design because the different system characteristics have a considerable effect on the architectural design, both functionally and aesthetically, and because the seismic system plays the major role in determining the seismic performance of the building. For example, if shear walls are chosen as the seismic force-resisting system, the building planning must be able to accept a pattern of permanent structural walls with limited openings that run uninterrupted through every floor from roof to foundation.

### 5.2.2 Diaphragms—the Horizontal Resistance System

The term "diaphragm" is used to identify horizontal-resistance members that transfer lateral forces between vertical-resistance elements (shear walls or frames). The diaphragms are generally provided by the floor and roof elements of the building; sometimes, however, horizontal bracing systems independent of the roof or floor structure serve as diaphrams. The diaphragm is an important element in the entire seismic resistance system (Figure 5-3).

The diaphragm can be visualized as a wide horizontal beam with components at its edges, termed **chords**, designed to resist tension and compression: chords are similar to the flanges of a vertical beam (Figure 5-3A)

A diaphragm that forms part of a resistant system may act either in a **flexible** or **rigid** manner, depending partly on its size (the area between enclosing resistance elements or stiffening beams) and also on its material. The flexibility of the diaphragm, relative to the shear walls whose

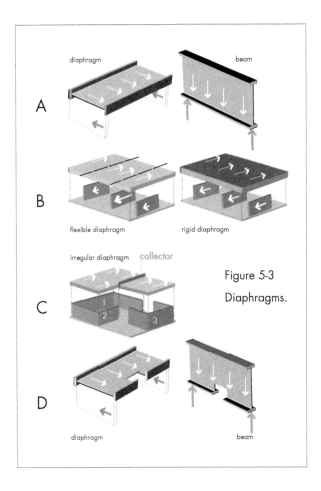

Figure 5-3

Diaphragms.

forces it is transmitting, also has a major influence on the nature and magnitude of those forces. With flexible diaphragms made of wood or steel decking without concrete, walls take loads according to tributary areas (if mass is evenly distributed). With rigid diaphragms (usually concrete slabs), walls share the loads in proportion to their stiffness (figure 5-3B).

**Collectors**, also called **drag struts** or **ties**, are diaphragm framing members that "collect" or "drag" diaphragm shear forces from laterally unsupported areas to vertical resisting elements (Figure 5-3C).

Floors and roofs have to be penetrated by staircases, elevator and duct shafts, skylights, and atria. The size and location of these penetrations are critical to the effectiveness of the diaphragm. The reason for this is not hard to see when the diaphragm is visualized as a beam. For example, it can be seen that openings cut in the tension flange of a beam will seriously weaken its load carrying capacity. In a vertical load-bearing situation, a penetration through a beam flange would occur in either a tensile or compressive region. In a lateral load system, the hole would be in a region of both tension and compression, since the loading alternates rapidly in direction (Figure 5-3D).

### 5.2.3 Optimizing the Structural/Architectural Configuration

Figure 5-4 shows the application of the three basic seismic systems to a model structural/architectural configuration that has been designed for near optimum seismic performance. The figure also explains the particular characteristics that are seismically desirable.

Building attributes:

○ **Continuous load path.**
Uniform loading of structural elements and no stress concentrations.

○ **Low height-to base ratio**
Minimizes tendency to overturn.

○ **Equal floor heights**
Equalizes column or wall stiffness, no stress concentrations.

○ **Symmetrical plan shape**
Minimizes torsion.

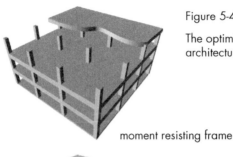

Figure 5-4

The optimized structural/architectural configuration.

moment resisting frame

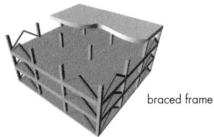

braced frame

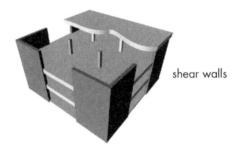

shear walls

◯ **Identical resistance on both axes**
Eliminates eccentricity between the centers of mass and resistance and provides balanced resistance in all directions, thus minimizing torsion.

◯ **Identical vertical resistance**
No concentrations of strength or weakness.

◯ **Uniform section and elevations**
Minimizes stress concentrations.

◯ **Seismic resisting elements at perimeter**
Maximum torsional resistance.

○ **Short spans**
Low unit stress in members, multiple columns provide redundancy -loads can be redistributed if some columns are lost.

○ **No cantilevers**
Reduced vulnerability to vertical accelerations.

○ **No openings in diaphragms(floors and roof)**
Ensures direct transfer of lateral forces to the resistant elements.

In the model design shown in Figure 5-4, the lateral force resisting elements are placed on the perimeter of the building, which is the most effective location; the reasons for this are noted in the text. This location also provides the maximum freedom for interior space planning. In a large building, resistant elements may also be required in the interior.

Since ground motion is essentially random in direction, the resistance system must protect against shaking in all directions. In a rectilinear plan building such as this, the resistance elements are most effective when placed on the two major axes of the building in a symmetrical arrangement that provides balanced resistance. A square plan, as shown here, provides for a near perfectly balanced system.

Considered purely as architecture, this little building is quite acceptable, and would be simple and economical to construct. Depending on its exterior treatment - its materials, and the care and refinement with which they are disposed-- it could range from a very economical functional building to an elegant architectural jewel. It is not a complete building, of course, because stairs, elevators, etc., must be added, and the building is not spatially interesting. However, its interior could be configured with nonstructural components to provide almost any quality of room that was desired, with the exception of unusual spatial volumes such as spaces more than one story in height.

In seismic terms, engineers refer to this design as a **regular** building. As the building characteristics deviate from this model, the building becomes increasingly **irregular**. It is these irregularities, for the most part created by the architectural design, that affect the building's seismic performance. **Indeed many engineers believe that it is these architectural irregularities that contribute primarily to poor seismic performance and occasional failure.**

## 5.3 THE EFFECTS OF CONFIGURATION IRREGULARITY

Configuration irregularity is largely responsible for two undesirable conditions-stress concentrations and torsion. These conditions often occur concurrently.

### 5.3.1 Stress Concentrations

Irregularities tend to create abrupt changes in strength or stiffness that may concentrate forces in an undesirable way. Although the overall design lateral force is usually determined by calculations based on seismic code requirements, the way in which this force is distributed throughout the structure is determined by the building configuration.

**Stress concentration** occurs when large forces are concentrated at one or a few elements of the building, such as a particular set of beams, columns, or walls. These few members may fail and, by a chain reaction, damage or even bring down the whole building. Because, as discussed in Section 4.10.2, forces are attracted to the stiffer elements of the building, these will be locations of stress concentration.

Stress concentrations can be created by both horizontal and vertical stiffness irregularities. The short-column phenomenon discussed in Section 4.10.2 and shown in Figure 4-14 is an example of stress concentration created by vertical dimensional irregularity in the building design. In plan, a configuration that is most likely to produce stress concentrations features **re-entrant corners:** buildings with plan forms such as an L or a T.) A discussion of the re-entrant corner configuration will be found in Section 5.5.4.

The vertical irregularity of the **soft or weak story** types can produce dangerous stress concentrations along the plane of discontinuity. Soft and weak stories are discussed in Section 5.5.1.

### 5.3.2 Torsion

Configuration irregularities in plan may cause torsional forces to develop, which contribute a significant element of uncertainty to an analysis of building resistance, and are perhaps the most frequent cause of structural failure.

As described in Section 4.11 and shown in Figure 4-17, torsional forces are created in a building by eccentricity between the center of mass and the center of resistance. This eccentricity originates either in the lack of symmetry in the arrangement of the perimeter-resistant elements as discussed in Section 5.5.3., or in the plan configuration of the building, as in the re-entrant-corner forms discussed in Section 5.5.4.

## 5.4 CONFIGURATION IRREGULARITY IN THE SEISMIC CODE

Many of the configuration conditions that present seismic problems were identified by observers early in the twentieth century. However, the configuration problem was first defined for code purposes in the 1975 *Commentary to the Strucural Engineers Association of California* (SEAOC) *Recommended Lateral Force Requirements* (commonly called the SEAOC Blue Book). In this section over twenty specific types of "irregular structures or framing systems" were noted as examples of designs that should involve further analysis and dynamic consideration, rather than the use of the simple equivalent static force method in unmodified form. These irregularities vary in importance in their effect, and their influence also varies in degree, depending on which particular irregularity is present. Thus, while in an extreme form the re-entrant corner is a serious plan irregularity, in a lesser form it may have little or no significance. The determination of the point at which a given irregularity becomes serious was left up to the judgment of the engineer.

Because of the belief that this approach was ineffective, in the 1988 codes a list of six horizontal (plan) and six vertical (section and elevation) irregularities was provided that, with minor changes, is still in today's codes. This list also stipulated dimensional or other characteristics that established whether the irregularity was serious enough to require regulation, and also provided the provisions that must be met in order to meet the code. Of the 12 irregularities shown, all except one are configuration irregularities; the one exception refers to asymmetrical location of mass within the building. The irregularities are shown in Figures 5.5 and 5.6. The code provides only descriptions of these conditions; the diagrams are added in this publication to illustrate each condition by showing how it would modify our optimized configuration, and to also illustrate the failure pattern that is created by the irregularity.

For the most part, code provisions seek to discourage irregularity in design by imposing penalties, which are of three types:

○ Requiring increased design forces.

○ Requiring a more advanced (and expensive) analysis procedure.

○ Disallowing extreme soft stories and extreme torsional imbalance in high seismic zones.

It should be noted that the code provisions treat the symptoms of irregularity, rather than the cause. The irregularity is still allowed to exist; the hope is that the penalties will be sufficient to cause the designers to eliminate the irregularities. Increasing the design forces or improving the analysis to provide better information does not, in itself, solve the problem. The problem must be solved by design.

The code-defined irregularities shown in Figures 5-5 and 5-6 serve as a checklist for ascertaining the possibility of configuration problems. Four of the more serious configuration conditions that are clearly architectural in origin are described in more detail in the sections below. In addition, some conceptual suggestions for their solution are also provided, as it may not be possible totally to eliminate an undesirable configuration.

## 5.5 FOUR SERIOUS CONFIGURATION CONDITIONS

Four configuration conditions (two vertical and two in plan) that originate in the architectural design and that have the potential to seriously impact seismic performance are:

○ Soft and weak stories

○ Discontinuous shear walls

○ Variations in perimeter strength and stiffness

○ Reentrant corners

| | performance | code remedies |
|---|---|---|
| | **P1  Torsional Irregularity: Unbalanced Resistance** | |
| | Localized damage. Collapse mechanism in extreme instances. | Modal Analysis, +65 foot high in SDC D,E,F. 25% increase to diaphragm connection design forces. Amplified forces to max of X3. |
| | **P2  Re-entrant Corners** | |
| | Local damage to diaphragm and attached elements. Collapse mechanism in extreme instances in large buildings. | 25% increase in diaphragm connection design forces. |
| | **P3  Diaphragm Eccentricity and Cutouts** | |
| | Localized structural damage. | 25% increase in diaphragm connection design forces. |
| | **P4  Nonparallel Lateral Force-Resisting System** | |
| | Leads to torsion and instability, localised damage. | Combine 100% and 30% of forces in 2 directions, use maximum. |
| | **P5  Out-of-Plane Offsets: Discontinuous Shearwalls** | |
| | Collapse mechanism in extreme circumstances. | Modal Analysis, +65 foot high in SDC D,E,F. 25% increase to diaphragm connection design forces. |

**resulting failure patterns**

**plan conditions**

SEISMIC ISSUES IN ARCHITECTURAL DESIGN

Figure 5-6: Vertical Irregularities (based on IBC, Section 1616.5.2).

| code remedies | performance | resulting failure patterns | vertical conditions |
|---|---|---|---|
| **V1 Stiffness Irregularity: Soft Story** | | | |
| Modal Analysis, +65 feet high in SCD D,E,F. Extreme case not permitted in seismic use groups E and F. | Common collapse mechanism. Death and much damage in Northridge earthquake. | | |
| **V2 Weight/Mass Irregularity** | | | |
| Modal Analysis, +65 foot high in SDC D,E,F. | Collapse mechanism in extreme circumstances. | | |
| **V3 Vertical Geometric Irregularity** | | | |
| Modal Analysis, +65 foot high in SDC D,E,F. | Localized structural damage. | | |
| **V4 In-Plane Irregularity in Vertical Lateral Force System** | | | |
| Modal Analysis, +65 foot high is SDC D, E, F. 25% increase to diaphragm connection design force. Supporting members designed for increased forces. | Localized structural damage. | | |
| **V5 Capacity Discontinuity: Weak Story** | | | |
| Modal Analysis, +65 foot high in SDC D,E,F. | Collapse mechanism in extreme circumstances | | |

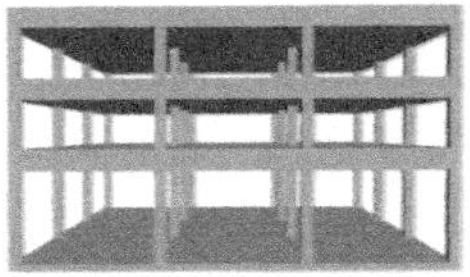

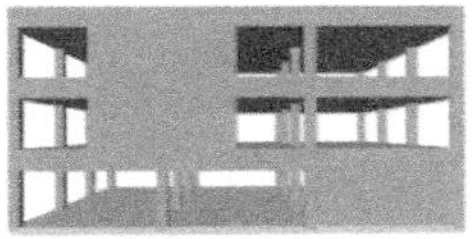

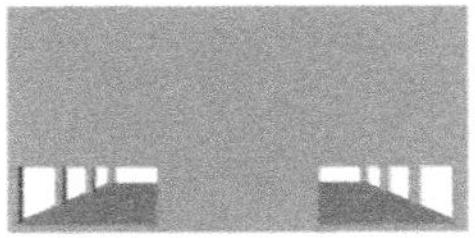

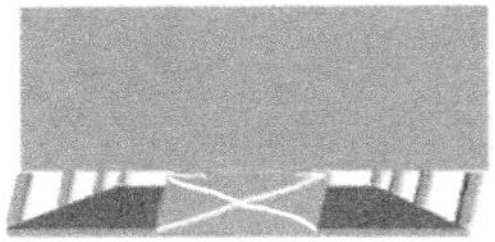

### 5.5.1 Soft and Weak Stories (Code Irregularities Types V1 and V5)

● The problem and the types of condition

The most prominent of the problems caused by severe stress concentration is that of the "soft" story. The term has commonly been applied to buildings whose ground-level story is less stiff than those above. The building code distinguishes between "soft" and "weak" stories. Soft stories are less stiff, or more flexible, than the story above; weak stories have less strength. A soft or weak story at any height creates a problem, but since the cumulative loads are greatest towards the base of the building, a discontinuity between the first and second floor tends to result in the most serious condition.

The way in which severe stress concentration is caused at the top of the first floor is shown in the diagram sequence in Figure 5-7. Normal drift under earthquake forces that is distributed equally among the upper floors is shown in Figure 5-7A. With a soft story, almost all the drift occurs in the first floor, and stress concentrates at the second-floor connections (Figure 5-7B). This concentration overstresses the joints along the second floor line, leading to distortion or collapse (Figure 5-7C).

Figure 5-7

The soft first story failure mechanism.

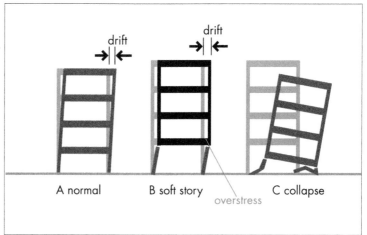

SEISMIC ISSUES IN ARCHITECTURAL DESIGN

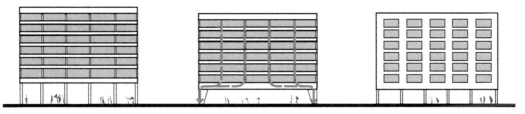

A flexible first floor   B discontinuity: indirect load path   C heavy superstructure

Figure 5-8: Three types of soft first story.

Three typical conditions create a soft first story (Figure 5-8). The first condition (Figure 5-8A) is where the vertical structure between the first and second floor is significantly more flexible than that of the upper floors. (The seismic code provides numerical values to evaluate whether a soft-story condition exists). This discontinuity most commonly occurs in a frame structure in which the first floor height is significantly taller than those above, so that the cube law results in a large discrepancy in stiffness (see Section 4.10.2 and Figure 4-13).

The second form of soft story (Figure 5-B) is created by a common design concept in which some of the vertical framing elements do not continue to the foundation, but rather are terminated at the second floor to increase the openness at ground level. This condition creates a discontinuous load path that results in an abrupt change in stiffness and strength at the plane of change.

Finally, the soft story may be created by an open first floor that supports heavy structural or nonstructural walls above (Figure 5-8C). This situation is most serious when the walls above are shear walls acting as major lateral force-resisting elements. This condition is discussed in Section 5.5.2, since it represents an important special case of the weak- and soft-story problem.

Figure 5-9 shows the Northridge Meadows apartment building after the Northridge (Los Angeles) earthquake of 1994. In this building, most of the first floor was left open for car parking, resulting in both a weak and flexible first floor. The shear capacity of the first-floor columns and the few walls of this large wood frame structure were quite inadequate, and led to complete collapse and 16 deaths.

Figure 5-9

Northridge Meadows apartments, Northridge earthquake , 1994.

Figure 5-10 shows another apartment house in Northridge in which two stories of wood frame construction were supported on a precast concrete frame. The frame collapsed completely. Fortunately there were no ground floor apartments, so the residents, though severely shaken, were uninjured.

Figure 5-10

Apartment building, Northridge earthquake, 1994. The first floor of this three-story apartment has disappeared.

● Solutions

The best solution to the soft and weak story problem is to avoid the discontinuity through architectural design. There may, however, be good programmatic reasons why the first floor should be more open or higher than the upper floors. In these cases, careful architectural/structural design must be employed to reduce the discontinuity. Some conceptual methods for doing this are shown in Figure 5-11.

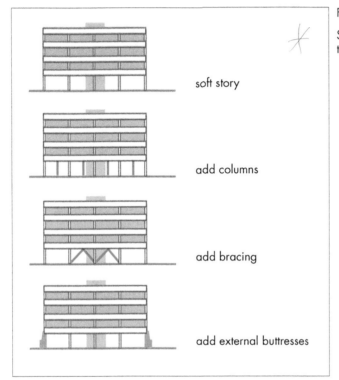

soft story

add columns

add bracing

add external buttresses

Figure 5-11

Some conceptual solutions to the soft first story.

Not all buildings that show slender columns and high first floors are soft stories. For a soft story to exist, the flexible columns must be the main lateral force-resistant system.

Designers sometimes create a soft-story condition in the effort to create a delicate, elegant appearance at the base of a building. Skillful structural/architectural design can achieve this effect without compromising the structure, as shown in Figure 5-12. The building shown is a 21-story apartment house on the beach in Vina del Mar, Chile. This building was unscathed in the strong Chilean earthquake of 1985.

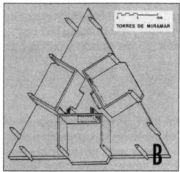

Figure 5-12: This apartment house appears to have a soft first story (Figure 5-12A), but the lateral force-resisting system is a strong internal shear wall box, in which the shear walls act as party walls between the dwelling units (Figure 5-12B). The architect achieved a light and elegant appearance, and the engineer enjoyed an optimum and economical structure.

### 5.5.2 Discontinuous Shear Walls (Code Type Irregularity V5)

● The problem and the types of condition

When shear walls form the main lateral resistant elements of a structure, and there is not a continuous load path through the walls from roof to foundation, the result can be serious overstressing at the points of discontinuity. This discontinuous shear wall condition represents a special, but common, case of the "soft" first-story problem.

The discontinuous shear wall is a fundamental design contradiction: the purpose of a shear wall is to collect diaphragm loads at each floor and transmit them as directly and efficiently as possible to the foundation. To interrupt this load path is undesirable; to interrupt it at its base, where the shear forces are greatest, is a major error. Thus the discontinuous shear wall that terminates at the second floor represents a "worst case" of the soft first-floor condition. A discontinuity in vertical stiffness and strength leads to a concentration of stresses, and the story that must hold up all the rest of the stories in a building should be the last, rather than the first, element to be sacrificed.

Olive View Hospital, which was severely damaged in the 1971 San Fernando, California, earthquake, represents an extreme form of the discontinuous shear wall problem. The general vertical configuration of the main building was a "soft" two-story layer of rigid frames on which was supported a four story (five, counting penthouse) stiff shear wall-plus-frame structure (Figures 5-13, 5-14). The second floor extends out to form a large plaza. Severe damage occurred in the soft story portion. The upper stories moved as a unit, and moved so much that the columns at ground level could not accommodate such a high displacement between their bases and tops, and hence failed. The largest amount by which a column was left permanently out-of-plumb was 2 feet 6 inches (Figure 5-15). The building did not collapse, but two occupants in intensive care and a maintenance person working outside the building were killed.

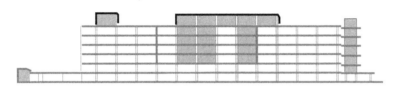

Figure 5-13: Long section, Olive View Hospital. Note that the shear walls stop at the third floor.

Figure 5-14: Cross section, Olive View hospital, showing the second-floor plaza and the discontinuous shear wall.

plaza level

Figure 5-15: Olive View hospital, San Fernando earthquake, 1971, showing the extreme deformation of the columns above the plaza level.

● Solutions

The solution to the problem of the discontinuous shear wall is unequivo-
cally to eliminate the condition. To do this may create architectural
problems of planning or circulation or image. If this is so, it indicates
that the decision to use shear walls as resistant elements was wrong from
the inception of the design. If the decision is made to use shear walls,
then their presence must be recognized from the beginning of schematic
design, and their size and location made the subject of careful architectural and engineering coordination early.

### 5.5.3 Variations in Perimeter Strength and Stiffness (Code Type P1)

● The problem and the types of condition

As discussed in Section 4.11, this problem may occur in buildings whose
configuration is geometrically regular and symmetrical, but nonetheless
irregular for seismic design purposes.

A building's seismic behavior is strongly influenced by the nature of
the perimeter design. If there is wide variation in strength and stiffness
around the perimeter, the center of mass will not coincide with the
center of resistance, and torsional forces will tend to cause the building
to rotate around the center of resistance.

Figure 5-16: Left, the building after the earthquake. Right, typical floor
plan showing the Center of Mass (CM), Center of Resistance (CR), and
Eccentricity (e) along the two axes.  PHOTO SOURCE: EERI

Figure 5-16 shows an apartment house in Viña del Mar, Chile, following the earthquake of 1985. The city is an ocean resort, and beach-front apartments are designed with open frontage facing the beach. This small seven-story condominium building had only three apartments per floor, with the service areas and elevator concentrated to the rear and surrounded by reinforced concrete walls that provided the seismic resistance. The lack of balance in resistance was such that the building rotated around its center of resistance, tilted sharply, and nearly collapsed. The building was subsequently demolished.

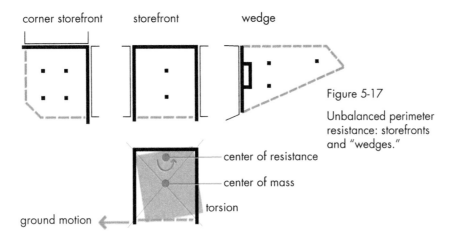

Figure 5-17

Unbalanced perimeter resistance: storefronts and "wedges."

A common instance of an unbalanced perimeter is that of open-front design in buildings, such as fire stations and motor maintenance shops in which it is necessary to provide large doors for the passage of vehicles. Stores, individually or as a group in a shopping mall, are often designed as boxes with three solid sides and an open glazed front (Figure 5-17).

The large imbalance in perimeter strength and stiffness results in large torsional forces. Large buildings, such as department stores, that have unbalanced resistance on a number of floors to provide large window areas for display are also common. A classic case of damage to a large store with an unbalanced-perimeter resistance condition was that of the Penney's store in the Alaska earthquake of 1964 (Figure 5-18).

---

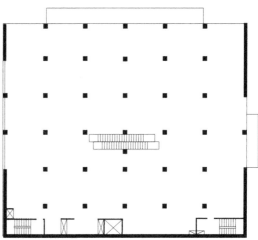

Figure 5-18: Penney's store, Anchorage, Alaska, earthquake, 1964. Left: Damage to the store: loss of perimeter precast panels caused two deaths. Right: Second-floor plan, showing unbalanced perimeter resistance. SOURCE: JAMES L. STRATTA

● Solutions

The solution to this problem is to reduce the possibility of torsion by endeavoring to balance the resistance around the perimeter. The example shown is that of the store front. A number of alternative design strategies can be employed that could also be used for the other building type conditions noted (Figure 5-19).

The first strategy is to design a frame structure of approximately equal strength and stiffness for the entire perimeter. The opaque portion of the perimeter can be constructed of nonstructural cladding, designed so that it does not affect the seismic performance of the frame. This can be done either by using lightweight cladding or by ensuring that heavy materials, such as concrete or masonry, are isolated from the frame (Figure 5-19A).

A second approach is to increase the stiffness of the open facades by adding sufficient shear walls, at or near the open face, designed to approach the resistance provided by the other walls (Figure 5-19B).

SEISMIC ISSUES IN ARCHITECTURAL DESIGN

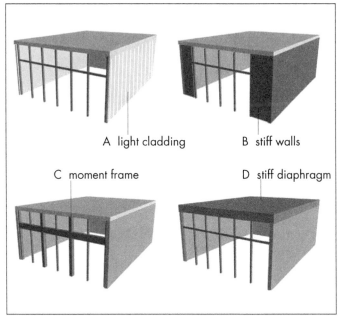

A  light cladding          B  stiff walls

C  moment frame          D  stiff diaphragm

A third solution is to use a strong moment resisting or braced frame at
the open front, which approaches the solid wall in stiffness. The ability
to do this will depend on the size of the facades; a long steel frame can
never approach a long concrete wall in stiffness. This is, however, a good
solution for wood frame structures, such as small apartment buildings,
or motels with ground floor garage areas, or small store fronts, because
even a comparatively long steel frame can be made as stiff as plywood
shear walls (Figure 5-19C).

The possibility of torsion may be accepted and the structure designed to
have the capacity to resist it, through a combination of moment frames,
shear walls,) and diaphragm action. This solution will apply only to relatively small structures with stiff diaphragms designed in such a way that
they can accommodate considerable eccentric loading (Figure 5-19D).

Manufacturers have recently produced prefabricated metal shear walls,
with high shear values, that can be incorporated in residential wood
frame structures to solve the house-over-garage problem.

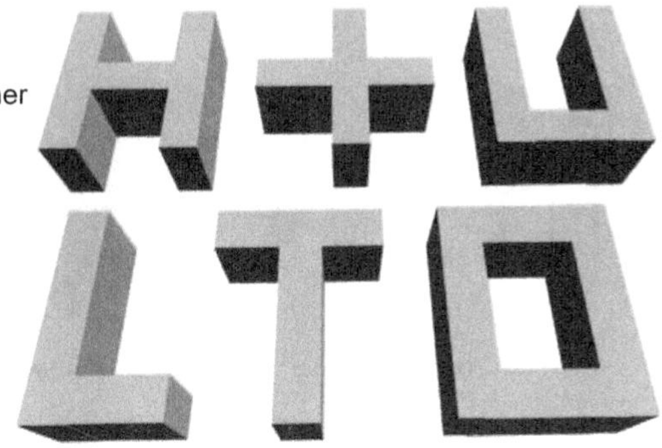

Figure 5-20

Re-entrant corner plan forms.

## 5.5.4 Re-entrant Corners (Code Type Irregularitiy H5)

● The problem and the types of condition

The re-entrant corner is the common characteristic of building forms that, in plan, assume the shape of an L, T, H, etc., or a combination of these shapes (Figure 5-20).

There are two problems created by these shapes. The first is that they tend to produce differential motions between different wings of the building that, because of stiff elements that tend to be located in this region, result in local stress concentrations at the re-entrant corner, or "notch".

The second problem of this form is torsion. Which is caused because the center of mass and the center of rigidity in this form cannot geo-metrically coincide for all possible earthquake directions. The result is rotation. The resulting forces are very difficult to analyze and predict. Figure 5-21 shows the problems with the re-entrant-corner form. The stress concentration at the "notch" and the torsional effects are interre-lated. The magnitude of the forces and the severity of the problems will depend on:

○  The characteristics of the ground motion

○  The mass of the building

○  The type of  structural systems

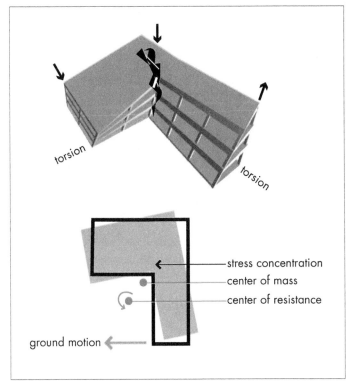

Figure 5-21

Re-entrant corner plan forms.

stress concentration
center of mass
center of resistance
ground motion

○ The length of the wings and their aspect ratios (length to width proportion)

○ The height of the wings and their height/depth ratios

Figure 5-22 shows West Anchorage High School, Alaska, after the 1964 earthquake. The photo shows damage to the notch of this splayed L-shape building.   Note that the heavy walls have attracted large forces. A short column effect is visible at the column between the two bottom windows which have suffered classic X –shaped shear-failure cracking and the damage at the top where this highly stressed region has been weakened by the insertion of windows.

Re-entrant corner plan forms are a most useful set of building shapes for urban sites, particularly for residential apartments and hotels, which enable large plan areas to be accommodated in relatively compact form, yet still provide a high percentage of perimeter rooms with access to air and light.

Figure 5-22: West Anchorage High School, Alaska earthquake, 1964. Stress concentration at the notch of this shallow L-shaped building damaged the concrete roof diaphragm.

SOURCE: NATIONAL INFORMATION SERVICE FOR EARTHQUAKE ENGINEERING, UNIVERSITY OF CALIFORNIA, BERKELEY.

These configurations are so common and familiar that the fact that they represent one of the most difficult problem areas in seismic design may seem surprising. Examples of damage to re-entrant-corner type buildings are common, and this problem was one of the first to be identified by observers.

The courtyard form, most appropriate for hotels and apartment houses in tight urban sites, has always been useful; in its most modern form, the courtyard sometimes becomes a glass-enclosed atrium, but the structural form is the same.

● Solutions

There are two basic alternative approaches to the problem of re-entrant-corner forms: structurally to separate the building into simpler shapes, or to tie the building together more strongly with elements positioned to provide a more balanced resistance (Figure 5-23). The latter solution applies only to smaller buildings.

SEISMIC ISSUES IN ARCHITECTURAL DESIGN

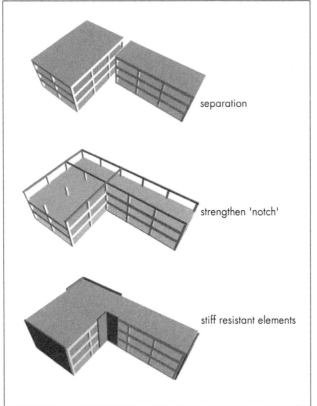

Figure 5-23

Solutions for the re-entrant-corner condition.

separation

strengthen 'notch'

stiff resistant elements

Once the decision is made to use separation joints, they must be designed and constructed correctly to achieve the original intent. Structurally separated entities of a building must be fully capable of resisting vertical and lateral forces on their own, and their individual configurations must be balanced horizontally and vertically.

To design a separation joint, the maximum drift of the two units must be calculated by the structural consultant. The worst case is when the two individual structures would lean toward each other simultaneously; and hence the sum of the dimension of the separation space must allow for the sum of the building deflections.

Several considerations arise if it is decided to dispense with the separation joint and tie the building together. Collectors at the intersection

Figure 5-24

Relieving the stress on
a re-entrant corner by
using a splay.

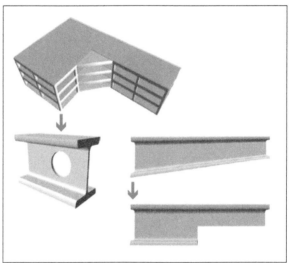

can transfer forces across the intersection area, but only if the design
allows for these beam-like members to extend straight across without interruption. If they can be accommodated, full-height continuous walls in
the same locations are even more effective. Since the portion of the wing
which typically distorts the most is the free end, it is desirable to place
stiffening elements at that location.

The use of splayed rather than right angle re-entrant corners lessens the
stress concentration at the notch (Figure 5-24). This is analogous to the
way a rounded hole in a steel plate creates less stress concentration than
a rectangular hole, or the way a tapered beam is structurally more desirable than an abruptly notched one.

## 5.6 OTHER ARCHITECTURAL/STRUCTURAL ISSUES

### 5.6.1 Overturning: Why Buildings Fall Down, Not Over

Although building mass or weight was discussed as part of the $F = MA$
equation for determining the horizontal forces, there is another way in
which the building's weight may act under earthquake forces to overload
the building and cause damage or even collapse.

Vertical members such as columns or walls may fail by buckling when the mass of the building exerts its gravity force on a member distorted or moved out of plumb by the lateral forces. This phenomenon is known by engineers as the **P-e** or **P-delta** effect, where P is the gravity force or weight, and "e" or "delta" is the eccentricity or the extent to which the force is offset. All objects that overturn do so as a result of this phenomenon (Figure 5-25).

The geometrical proportions of the building also may have a great influence on whether the P-delta effect will pose a problem, since a tall, slender building is much more likely to be subject to overturning forces than a low, squat one. It should be noted, however, that if the lateral resistance is provided by shear walls, it is the proportions of the shear walls that are significant rather than those of the building as a whole.

However, in earthquakes, buildings seldom overturn, because structures are not homogeneous but rather are composed of many elements connected together; the earthquake forces will pull the components apart, and the building will fall down, not over. Strong, homogeneous structures such as filing cabinets, however, will fall over. A rare example of a large steel-frame building collapse is that of the Piño Suarez apartments in the Mexico City earthquake of 1985. Of the three nearly identical buildings, one collapsed, one was severely damaged, and the third

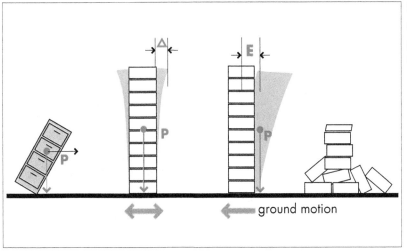

ground motion

Figure 5-25

Why buildings generally fall down, not over.

Figure 5-26

Piño Suarez apartments, Mexico City, 1985.

SOURCE: NIST

suffered moderate damage. The structures had asymmetrical lateral bracing at their perimeters, and the steel frames were poorly detailed and buckled (Figure 5-26).

The collapse of the Cypress Freeway in Oakland, California, in the Loma Prieta earthquake (though a viaduct rather than building) was a rare example of a low-rise structural collapse (Figure 5-27),

## 5.6.2 Perforated Shear Walls

Another undesirable condition is when a shear wall is perforated by aligned openings for doors , windows and the like, so that its integrity may be compromised. Careful analysis is necessary to ensure that a continuous load path remains without a significant loss of horizontal shear capacity. Some types of perforated shear wall with unaligned openings have performed well (Figure 5-28).

Figure 5-27

Collapse of large two-story section of the Cypress Freeway, San Francisco, Loma Prieta earthquake, 1989.

Figure 5-28

Shear wall perforated by large opening (at bottom right-hand corner).

### 5.6.3 Strong Beam, Weak Column

Structures are commonly designed so that under severe shaking, the beams will fail before the columns. This reduces the possibility of complete collapse. The short-column effect, discussed in Section 4.10.2, is analogous to a weak-column strong-beam condition, which is sometimes produced inadvertently when strong or stiff nonstructural spandrel members are inserted between columns. The parking structure shown in Figure 5-29 suffered strong-beam weak-column failure in the Whittier, California, earthquake of 1987.

### 5.6.4 Setbacks and Planes of Weakness

Vertical setbacks can introduce discontinuities, particularly if columns or walls are offset at the plane of the setback. A horizontal plane of weakness can be created by the placement of windows or other openings that may lead to failure, as in this building in the Kobe, Japan, earthquake of 1995 (Figure 5-30).

Figure 5-29: Damaged parking structure, Whittier Narrows (Los Angeles) earthquake, 1987. The deep spandrels create a strong-beam, weak-column condition.

## 5.7 IRREGULAR CONFIGURATIONS: A TWENTIETH CENTURY PROBLEM

The foregoing discussion has identified "irregular" architectural/structural forms that can contribute to building damage or even collapse. These irregularities are present in many existing buildings, and the ways in which they affect seismic performance need to be understood by building designers so that dangerous conditions are not created. The ir-

Figure 5-30

Damaged building, Kobe earthquake, Japan, 1995.

SEISMIC ISSUES IN ARCHITECTURAL DESIGN

regular-configuration problem was made possible by nineteenth-century structural technology and created by twentieth-century architectural design.

### 5.7.1 A New Vernacular: the International Style and its Seismic Implications

The innovation of the steel and reinforced concrete frame at the end of the nineteenth century enabled buildings to be freed from the restrictions imposed by load-bearing masonry. However, until the early years of the twentieth century, western architectural design culture dictated a historical style even when totally new building types, such as railroad stations or skyscrapers, were conceived. The architectural forms used were all derived from the engineering imperatives of load-bearing masonry structure: these masonry-devised forms survived well into the twentieth century, even when buildings were supported by concealed steel frames, and arches had become stylistic decoration (Figure 5-31).

Figure 5-31

Early twentieth-century steel-frame buildings, Michigan Avenue, Chicago.

This historicism came under attack early in the century from a number of avant-garde architects, predominantly in Europe, who preached an anti-historical dogma in support of an architecture that they believed more fully represented the aspirations and technology of a new age. Later, this movement was termed the **International Style**.

This revolution in architectural aesthetics had many dimensions: aesthetic, technical, economic and political. One result was to give aesthetic validity to a highly economical, unadorned, rectilinear box for almost all building functions. The international style preached the aesthetic enjoyment of the delicacy and slenderness that the steel or concrete frame structure had made possible.

The prototype of the international style was exemplified in the Pavillon Suisse in Paris in 1930 (Figure 5-32).

Figure 5-32

The Pavillon Suisse, Le Corbusier, Paris, 1930: elevated on pilotis, use of a free plan, and curtain walls.

As architects and engineers began to exploit the aesthetics of the building frame, the seeds of seismic configuration problems were sown. In its earliest forms the style frequently created buildings that were close to our ideal seismic building configuration. However, the style often had a number of characteristics not present in earlier frame and masonry buildings that led to poor seismic performance. These were:

○ Elevation of the building on stilts or pilotis

This had attractive functional characteristics, such as the ability to introduce car parking under the building, or the building could be opened to the public and its visitors in ways that were not previously possible. It was attractive aesthetically: the building could appear to float airily above the ground.

However, without full understanding of the seismic implications of vertical structural discontinuity, designers often created soft and weak stories.

○ The free plan and elimination of interior-load bearing walls

Planning freedom was functionally efficient and aesthetically opened up new possibilities of light and space.

However, the replacement of masonry and tile partitions by frame and gypsum board greatly reduced the energy absorption capability of the building and increased its drift, leading to greater nonstructural damage and possible structural failure.

○ The great increase in exterior glazing and the invention of the light-weight curtain wall

The curtain wall was a significant feature of the new vernacular and was subject to continuous development and refinement. At one end, it became the most economical method of creating an exterior façade; at the other end it led to the apparently frameless glass walls and double-skin energy-efficient curtain walls of today. Like free interior planning, the light exterior cladding greatly reduced the energy-absorption capability of the building and increased its drift.

The post-World War II years saw worldwide explosive urban development, and the new aesthetic, because of its lack of ornamentation, simple forms, and emphasis on minimal structure, was very economical. This ensured its widespread adoption. Unfortunately, seismic design, particularly the need for ductility - as it related to the new, spare, framed buildings - was inadequately understood. Thus the aesthetics and economies of the international style in vogue from about the 50's to the 70's has left the world's cities with a legacy of poor seismic configurations that presents a serious problem in reducing the earthquake threat to our towns and cities.

Configuration irregularities often arise for sound planning or urban design reasons and are not necessarily the result of the designer's whim (or ignorance). The problem irregularities shown in Figures 5-5 and 5-6 represent structural/architectural errors that originate in the architectural design as the result of a perceived functional or aesthetic need. The errors can be avoided through design ingenuity, and mutual understanding and a willingness to negotiate design issues between the architect and engineer. The architect needs to understand the possible implications of the design, and the engineer needs to embrace the design objectives and participate in them creatively.

## 5.8 DESIGNING FOR PROBLEM AVOIDANCE

Regardless of building type, size, or function, it is clear that the attempt to encourage or enforce the use of regular configurations is frequently not going to succeed; the architect's search for original forms is very powerful. The evolution and recent trends in formal invention are shown in Figure 5-38 in Section 5.9.2.

The seismic code, as illustrated in Figures 5-5 and 5-6, is oriented towards "everyday" economical building and goes a modest route of imposing limited penalties on the use of irregular configurations in the form of increased design forces and, for larger buildings, the use of more advanced analytical methods; both these measures translate into cost penalties Only two irregularities are banned outright: extreme soft stories and extreme torsion in essential buildings in high seismic zones. This suggests a strategy that exploits the benefits of the "ideal" configuration but permits the architect to use irregular forms when they suit the design intentions.

### 5.8.1 Use of Regular Configurations

A design that has attributes of the ideal configuration should be used when:

○ The most economical design and construction is needed, including design and analysis for code conformance, simplicity of seismic detailing, and repetition of structural component sizes and placement conditions.

○ When best seismic performance for lowest cost is needed.

○ When maximum predictability of seismic performance is desired.

### 5.8.2 Designs for Irregular Configurations

When the design incorporates a number of irregularities the following procedures should be used:

○ A skilled seismic engineer who is sympathetic to the architect's design intentions should be employed as a co-designer from the outset of the design.

○ The architect should be aware of the implications of design irregularities and should have a feel for the likelihood of stress concentrations and torsional effects (both the cause and remedy of these conditions lie in the architectural/structural design, not in code provisions).

○ The architect should be prepared to accept structural forms or assemblies (such as increased size of columns and beams) that may modify the design character, and should be prepared to exploit these as part of the aesthetic language of the design rather than resisting them.

○ The architect and engineer should both employ ingenuity and imagination of their respective disciplines to reduce the effect of irregularities, or to achieve desired aesthetic qualities without compromising structural integrity.

○ Extreme irregularities may require extreme engineering solutions; these may be costly, but it is likely that a building with these conditions will be unusual and important enough to justify additional costs in materials, finishes, and systems.

○ A soft or weak story should never be used: this does not mean that high stories or varied story heights cannot be used, but rather that appropriate structural measures be taken to ensure balanced resistance.

## 5.9 BEYOND THE INTERNATIONAL STYLE: TOWARDS A SEISMIC ARCHITECTURE?

Most owners desire an economical and unobtrusive building that will satisfy the local planning department and look nice but not unusual. However, as noted above, the occasional aspiration for the architect to provide a distinctive image for the building is very powerful and is the source of continued evolution in architectural style and art. This thrust is allied to today's "marketing" demand for spectacular forms. The history of architecture shows that design innovation has its own life, fed by brilliant form-givers who provide prototypes that keep architecture alive and exciting as an art form. Thus, like economics, architectural design has its "supply- and demand-sides" that each reinforce one another.

The International Style still exists as a vernacular and can range from everyday economical buildings to refined symbols of prestige. But there are now many competing personal styles. Have the tenets of good seismic design played any role in determining their characteristics? Is it possible that future architectural stylistic trends might seek inspiration in seismic design as an aesthetic that matches the exigencies of physics and engineering with visual grace and intrigue?

### 5.9.1 The Architect's Search for Forms – Symbolic and Metaphorical

The aesthetic tenets of the International Style—particularly the metal/glass cubistic building—began to be seriously questioned by the mid-1970s. This questioning finally bore fruit in an architectural style known broadly as post-modern. Among other characteristics, post-modernism embraced:

○ The use of classical forms, such as arches, decorative columns, pitched roofs in nonstructural ways and generally in simplified variations of the original elements

○ The revival of surface decoration on buildings

○ A return to symmetry in configuration

In seismic terms, these changes in style were, if anything, beneficial. The return to classical forms and symmetry tended to result in regular structural/architectural configurations, and almost all of the decora-

Figure 5-33

Portland Building, Portland, OR. Architect: Michael Graves, 1982.

SEISMIC ISSUES IN ARCHITECTURAL DESIGN

tive elements were nonstructural. An early icon of post-modernism, the Portland, Oregon, office building, designed by Michael Graves (Figure 5-33) used an extremely simple and conservative structural system. Indeed, this building, which created a sensation when completed, has a structural/architectural configuration that is similar to the model shown in Figure 5-33. The sensation was all in the nonstructural surface treatments, some proposed exterior statues, and in its colors.

A conventionally engineered steel or concrete member that was supporting the building could be found inside every classical post-modern column. It is clear that an interest in seismic design or structure in general had no influence on the development of post-modernism; it was strictly an aesthetic and cultural movement.

At the same time that post-modernism was making historical architectural style legitimate again, another style began to flourish, to some extent in complete opposition. This style (originally christened "hi-tech") returned to the celebration of engineering and new industrial techniques and materials as the stuff of architecture. This style originated primarily in Europe, notably in England and France, and the influence of a few seminal works, such as the Pompidou Center in Paris (Figure 5-34).

Although seismic concerns had no influence on the origin and development of this style, it is relevant here because it revived an interest in exposing and celebrating structure as an aesthetic motif.

Figure 5-34

Pompidou Center, Paris, Architect: Piano and Rogers, 1976.

Post-modernism died a quick death as an avant-garde style, but it was important because it legitimized the use of exterior decoration and classically derived forms. These became common in commercial and institutional architecture (Figure 5-35). The notion of "decorating" the economical cube with inexpensive simplified historic or idiosyncratic nonstructural elements has become commonplace.

Figure 5-35

Post-modern influences, 2000.

At the same time, in much everyday commercial architecture, evolved forms of the International Style still predominate, to some extent also representing simplified (and more economical) forms of the high-tech style. Use of new lightweight materials such as glass fiber-reinforced concrete and metal-faced insulated panels has a beneficial effect in reducing earthquake forces on the building, though provision must be made for the effects of increased drift on nonstructural components or energy-dissipating devices used to control it.

## 5.9.2 New Architectural Prototypes Today

The importance of well-publicized designs by fashionable architects is that they create new prototypical forms. Architects are very responsive to form and design, and once a new idiom gains credence, practicing architects the world over begin to reproduce it. Today's New York corporate headquarters high-rise becomes tomorrow's suburban savings and loan office, as shown in Figures 5-36 and 5-37.

Today, however, unlike the era of the International Style and the adoption of "modern" architecture, there is no consensus on a set of appropriate forms. At present, spectacular architectural design is in fashion

Figure 5-36

United Nations Secretariat, New York, Architects: Wallace Harrison, Le Corbusier, Oscar Niemeyer, and Sven Markelius, 1950.

and sought after by municipalities, major corporations, and institutions. So, it is useful to look at today's cutting-edge architecture, because among it will be found the prototypes of the vernacular forms of the future.

Figure 5-38 shows the evolution of the architectural form of the high-rise building from the 1920s to today. There is a steady evolution in which the international style dominates the scenes from about 1945 to 1985. For a brief interlude, post-modern architecture is fashionable, in company with "high-tech". Towards the end of the century, architec-

Figure 5-37

Main street vernacular, anywhere, USA.

Figure 5-38:   The evolution of high-rise building form. The twentieth century was a period of evolution.

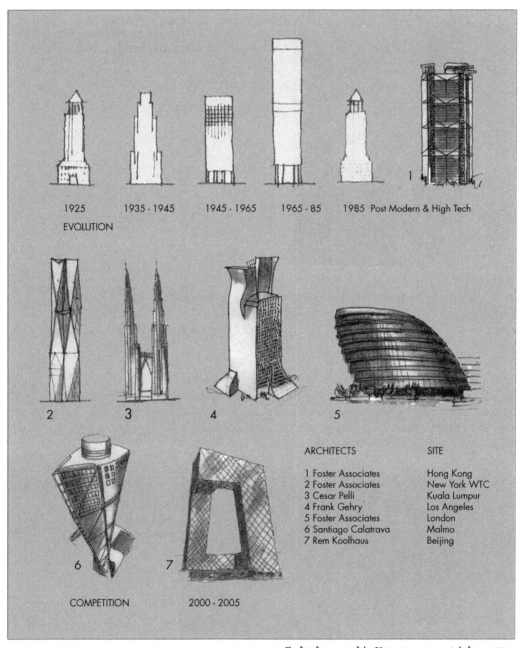

The first five years of the 21st century are a period of competition.

SEISMIC ISSUES IN ARCHITECTURAL DESIGN

tural forms become more personal and idiosyncratic, and evolution is replaced by competition. The first five years of the millenium have seen the emergence of a number of very personal styles, from the jagged forms of Liebskind to the warped surfaces of Gehry. The Foster office in London pursues its own in-house evolution of high-tech design.

In general, today's high-rise buildings remain vertical, and have direct load paths, and their exterior walls are reasonably planar. Some high-rise towers have achieved a modest non-verticality by the use of nonstructural components. A more recent development is that of the "torqued" tower, as in the Freedom Tower at the World Trade Center and Santiago Calatrava's "Turning Torso" tower in Malmo, Sweden, shown in Figure 5-38. For very tall buildings, it is claimed that these twisted forms play a role in reducing wind forces, besides their visual appeal, but their forms are not of significance seismically.

In lower buildings, where there is more freedom to invent forms than in the high rise, planning irregularities (and corresponding three-dimensional forms) are now fashionable that go far beyond the irregularities shown in Figure 5-6. Figure 5-39 shows the extraordinary range of plan forms for art museums conceived by four of today's most influential architects.

Highly fragmented facades now abound, serving as metaphors for the isolated and disconnected elements of modern society. Often-repeated design motifs include segmental, undulating, or barrel-vaulted roofs and canopies, and facades that change arbitrarily from metal and glass curtain wall to punched-in windows.

In all this ferment, there is much originality and imagination, and often high seriousness. It remains to be seen whether any of these forms become attractive to the typical practitioner and their more conservative clients; however, indications of the influence of some of these motifs can now be discerned in more commonplace buildings along the highways and in schools and universities (Figure 5-40).

One may question the extent to which architectural trends look as if they will increase or decrease the kinds of configuration irregularities that manifested themselves in the international style era. The answer appears

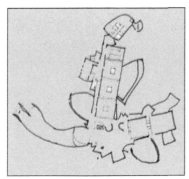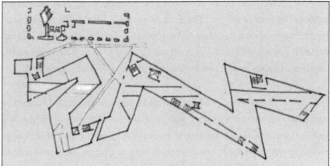

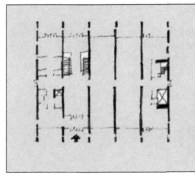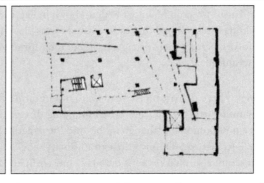

Figure 5-39: Planning variety: four plans of new museums. Top left, Guggenheim Museum, Bilbao, Spain, Architect Frank Gehry, 1998. Top right, Jewish Museum, Berlin, Architect: Daniel Liebskind, 1999. Bottom left, Rosenthal Center for the Arts, Cincinatti, Ohio, Architect: Zaha Hadid 2003, Bottom right, Nasher Sculpture Center, Dallas, Texas ,Architect : Renzo Piano Design Workshop, 2003.

Northern Spain is a low seismic zone. Cincinnati, Berlin, and Dallas are not subject to earthquakes.

Figure 5-40: The influence of prototypes: fragmented facades and tilted walls.

　　　　　　SEISMIC ISSUES IN ARCHITECTURAL DESIGN

to be that they will increase, because much new architecture is clearly conceived independently of structural concerns or in the spirit of theatrical set design, with the engineer in the role of an enabler rather than collaborator.

### 5.9.3 Towards an Earthquake Architecture

In the search for meaning in architecture that supersedes the era of International Style and the superficialities of fashion exemplified by much of post-modernism and after, perhaps architects and engineers in the seismic regions of the world might develop an "earthquake architecture". One approach is an architecture that expresses the elements necessary to provide seismic resistance in ways that would be of aesthetic interest and have meaning beyond mere decoration. Another approach is to use the earthquake as a metaphor for design.

### 5.9.4 Expressing the Lateral-Force Systems

For the low and midrise building, the only structural system that clearly expresses seismic resistance is the use of exposed bracing. There are historical precedents for this in the half-timbered wood structures of medieval Germany and England. This was a direct and simple way of bracing rather than an aesthetic expression, but now these buildings are much prized for their decorative appearance. Indeed, the "half-timbered" style has become widely adopted as an applied decorative element on U.S. architecture, though for the most part at a modest level of residential and commercial design.

Two powerful designs in the 1960s, both in the San Francisco Bay Area, used exposed seismic bracing as a strong aesthetic design motif. These were the Alcoa Office Building and the Oakland Colisem, both designed in the San Francisco office of Skidmore, Owings and Merrill (Figure 5-41).

In spite of these two influential designs and others that used exposed wind bracing, the subsequent general trend was to de-emphasize the presence of lateral-resistance systems. Architects felt that they conflicted with the desire for purity in geometric form, particularly in glass "box" architecture, and also possibly because of a psychological desire to deny the prevalence of earthquakes. However, in the last two decades it has become increasingly acceptable to expose lateral-bracing systems and enjoy their decorative but rational patterns (Figure 5-42).

Figure 5-41: Left: Alcoa Building, San Francisco, 1963. Right: Oakland Coliseum, 1960. Architect: Skidmore Owings and Merrill.

Figure 5-42: Exposed cross-bracing examples.

Top left; Pacific Shores Center, Redwood City, CA, Architects DES Architects & Engineers. Top right: Silicon Graphics, Mountain View, Architects: Studios Architects. Bottom left: Sports Arena, San Jose, Architects: Sink, Combs, Dethlefs, (All in California). Bottom right: Government Offices, Wanganui, New Zealand.

Figure 5-43: Left: Retrofitted student residences. Right: University Administration Building, Berkeley California, Architect: Hansen, Murakami and Eshima, Engineers; Degenkolb Engineers.

This new acceptability is probably due to boredom with the glass cube and the desire to find a meaningful way of adding interest to the façade without resorting to the applied decoration of post-modernism. In addition, greater understanding of the earthquake threat has led to realization that exposed bracing may add reassurance rather than alarm.

Exposed bracing is also used as an economical retrofit measure on buildings for which preservation of the façade appearance is not seen as important. A possible advantage of external bracing is that often the building occupants can continue to use the building during the retrofit work, which is a major economic benefit; however, see Chapter 8.5.3.1 for further discussion of this point. External bracing retrofits have also sometimes had the merit of adding visual interest to a number of dull 1960s rectilinear type facades (Figure 5-43).

The movement towards exposed seismic bracing has some parallels with the aesthetic movement of exposing buildings' mechanical systems. Designers who had become bored with expanses of white acoustical ceiling realized that mechanical systems, particularly when color-coded, were of great visual interest and also intrigued those who are fascinated by mechanical systems and devices. Another parallel with seismic design is

Figure 5-44: Elegantly expressed exposed bracing:
Left: University Administration Building, Berkeley
California, Architect: Hansen Murakami and Eshima.
Right: Millenium Bridge, London, 2000. Architect:
Foster Associates; Arup Engineers, Engineer.

that, when mechanical systems were exposed, their layout and detailing
had to be much more carefully designed and executed, from an aesthetic
viewpoint. In a similar way, exposed bracing has to be more sensitively
designed, and this has seen the development of some elegant design and
material usage (Figure 5-44).

New innovations, such as base isolation and energy absorbing devices,
have sometimes been exploited for aesthetics and reassurance. The designers of an early and ingenious base isolated building in New Zealand
(the Union House office building in Auckland) not only exposed its
braced-frame, but also made visible its motion-restraint system at its open
first-floor plaza (Figure 5-45).

Experiments in linking the rationality of structure to the poetics of
form and surface are shown in Figure 5-46, which shows two schemes
for advanced systems of perimeter bracing that, if exposed, are perhaps
livelier than conventional concealed bracing. The left hand figure shows
a 60 story structure with 10 story braced super frame units, restrained
by periodic two story moment frame clusters with hydraulic dampers.
The right-hand figure shows a 48 story moment frames with random
offset toggle hydraulic dampers. The apparent random character of the
bracing is based on the load patterns within the structure.

Figure 5-45: Left: Union House, Auckland, New Zealand. Right: detail of energy absorbing system. Architect: Warren and Mahoney; Engineer, Brian Wood

The intent is to exploit an interest in structural expression and its forms, and create a "code" that can be read by anyone that has a sense of how lateral forces operate and must be resisted.

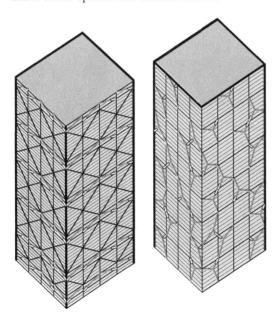

Figure 5-46: Left: 60-story structure with 10 story braced super frame units, restrained by periodic two story moment frame clusters with hydraulic dampers. Right: 48-story moment frames with random offset toggle hydraulic dampers.

### 5.9.5 The Earthquake as a Metaphor

A more theoretical use of the earthquake as a design inspiration is that of designing a building that reflects the earthquake problem indirectly, as a metaphor. This approach is rare, but has some interesting possibilities for certain building types, such as seismic engineering laboratories.

One of the few executed examples of this approach is the Nunotani Office Building in Tokyo. The architect, Peter Eisenman of New York, says that the building represents a metaphor for the waves of movement as earthquakes periodically compress and expand the plate structure of the region (Figure 5-47).

A listing of ideas for this metaphorical approach has been suggested as part of a student design project at the architecture school, Victoria University, New Zealand (Table 5.1). Figure 5-48 shows a student project in which damage is used as a metaphor, following the example of the Nunotani Building.

Figure 5-47

Nunotani Office building, Tokyo, Architect: Peter Eisenman 1998

The architect/artist Lebbeus Woods has created imaginary buildings in drawings of extraordinary beauty that explicitly use the representation of seismic forces as a theme (Figure 5-49).

In his project "Radical Reconstruction," Woods was inspired by the 1995 Kobe earthquake to explore the implications of building destruction. Of his many drawings and paintings inspired by San Francisco, Woods has written that these projects "explore the possibilities for an architecture that in its conception, construction and inhabitation comes into new and potentially creative relationships not only with the effects of earthquakes, but more critically with the wider nature of which they are a part".

The expression of seismic resistance and the metaphor of the earthquake could yet provide a rich creative field for a regional architecture that derives at least some of its aesthetic power from the creation of useful and delightful forms that also celebrate the demands of seismic forces and the way they are resisted.

Figure 5-48

Student project, damage as a metaphor.

Designer: L. Allen

Table 5-1: Potential design ideas listed under various headings

| Geology & Seismology | Construction Issues | General Concepts or Ideas not Specifically Related to | Other Earthquake Related Items |
|---|---|---|---|
| Seismic waves | Propping | Healing processes such as scabs that form after injury | Temporary buildings for disaster relief |
| Faulting | Tying elements together | External forces on a building | Seismographs |
| Earthquake-affected landforms | Post-earthquake ruins | Adaptability | Expression of structural action |
| Contrast between geologic and seismografic scale | Disassembly | Insecurity | Brittle behavior |
| | Seismic-resisting technology | Preparedness | Plastic behavior |
| | Contrast between gravity and lateral load-resisting structure | Engineer & architect relationship | |

Figure 5-49: Lebbeus Woods: detail of "San Francisco Project: inhabiting the quake WAVE house drawing", 1995. In this theoretical design, "the ball-jointed frames flex and re-flex in the quake: supple metal stems and leaves move in the seismic winds".

SOURCE: LEBBEUS WOODS, REDICAL RECONSTRUCTUION, PRINCETON ARCHITECTURAL PRESS, NEW YORK, 1997

## 5.10 CONCLUSION

This chapter has focused on basic seismic structural systems in relation to architectural configurations, and has looked at architectural design through a seismic "filter". This shows that many common and useful architectural forms are in conflict, with seismic design needs. To resolve these conflicts the architect needs to be more aware of the principles of seismic design, and the engineer needs to realize that architectural configurations are derived from many influences, both functional and aesthetic. The ultimate solution to these conflicts depends on the architect and engineer working together on building design from the outset of the project and engaging in knowledgeable negotiation.

Trends in architectural taste suggest that for the engineer to expect to convince the architect of some of the conventional virtues of seismic design, such as simplicity, symmetry and regularity, is only realistic for projects in which economy and reliable seismic performance are paramount objectives. When the architect and the client are looking for high-style design, the forms will probably be irregular, unsymmetrical, and fragmented. The wise and successful engineer will enjoy the challenges. New methods of analysis will help, but engineers must also continue to develop their own innate feeling for how buildings perform, and be able to visualize the interaction of configuration elements that are quite unfamiliar.

## 5.11 REFERENCES

Structural Engineers Association of California (SEAOC) *Blue Book*

International Code Council, *International Building Code*, Birmingham AL, 2003

Lebbeus Woods: *Radical Reconstruction*, Princeton Architectural Press, New York, NY, 1997

Andrew Charleson and Mark Taylor: *Earthquake Architecture Explorations, Proceedings*, 13thWorld Conference on Earthquake Engineering, Vancouver, BC 2004

Mark Taylor, Julieanna Preston and Andrew Charleson, *Moments of Resistance*, Archadia Press, Sydney, Australia, 2002

## 5.12 TO FIND OUT MORE

Christopher Arnold, *Architectural Considerations (chapter 6), The Seismic Design Handbook, Second Edition* ( Farzad Naeim, ed.) Kluver Academic Publishers, Norwell, MA 2001

Terence Riley and Guy Nordenson, *Tall Buildings*, The Museum of Modern Art, New York, NY, 2003

Sheila de Vallee, *Architecture for the Future*, Editions Pierre Terrail, Paris, 1996

Maggie Toy, ed. *Reaching for the Sky*, Architectural Design, London, 1995

Yukio Futagawa, ed, *GA Document*. A serial chronicle of modern architecture, A.D.A Edita, Tokyo, published periodically

Garcia, B, (ed.) *Earthquake Architecture*, Loft and HBI an imprint of Harper Collins International, New York, NY, 2000

Sandaker, B. N. and Eggen, A. P. *The Structural Basis of Architecture*, Phaiden Press Ltd., London, 1993

by Christine Theodoropolous

## 6.1 INTRODUCTION

This chapter presents an overview of the development of seismic design codes in the United States. It includes a discussion of concepts underlying performance-based seismic design and addresses how seismic provisions in current model building codes can inform architectural design decisions.

Readers can find an explanation of the basis for hazard maps that are used to define code-specified design earthquake parameters in Chapter 2. Chapter 3 discusses the seismic zonation principles used to generate seismic design regulations that restrict property use or require site-specific design approaches. Chapter 5 explains how building codes classify building configurations for seismic design purposes. Chapter 8 discusses the regulatory environment for seismic-design involving existing buildings.

Although building codes have evolved substantially in the last hundred years, they still reflect history in the way that they are organized and used. Code sections that originated as fire mitigation measures precede the sections containing the structural provisions that include seismic design requirements. Sections pertaining to modern building systems, such as plumbing and electrical systems, are published as separate volumes. In conventional building design practice, architects take primary responsibility for addressing fire-related provisions. Although the use of fire-resistive design consultants is becoming common in certain kinds of institutional and commercial building designs, architects using these consultants will still develop initial conceptual designs in accordance with the principles that underlie fire-resistive code measures.

The same is not commonly true in seismic design. Although building configuration and other design features, determined by the architect, are known to impact the structural and nonstructural performance of buildings in earthquakes, architects do not usually consult the sections of the code where the seismic design requirements are addressed. Codes governing seismic design were established in the second half of the twentieth century, primarily by structural engineers, and reflect the increasing specialization and disciplinary division between architectural and engineering practice. Few architects practicing today take responsibility for structural aspects of seismic design, and many consult with engineers on the design of nonstructural architectural elements that are regulated by seismic code provisions. However, architects routinely make design decisions that impact the use and interpretation of seismic design codes.

## 6.2 EARTHQUAKES AND CODE ACTION

Historically, seismic design provisions were added to codes in response to the lessons learned from earthquake damage. Although the evolution of technical understanding of building performance has guided the development of these provisions, code action has been driven primarily by political rather than technical advances. Communities with well-developed political mechanisms for addressing public safety have tended to pioneer code developments, but the long periods between damaging earthquakes have made it easy for communities to forget to follow through with efforts begun in the aftermath of disasters. In addition, the political and technical complexities inherent in extracting lessons from earthquakes have made it difficult to achieve consensus on appropriate code measures.

### 6.2.1 Early 20th Century

Despite the destruction of 27,000 buildings and fatalities estimated by the USGS (United States Geological Survey) and others to be between 700 and 2,100, the 1906 San Francisco earthquake did not stimulate an explicit code response. A wind-load requirement was implemented and was assumed to be sufficient to resist earthquake forces. Post-earthquake investigators reported that 80% to 95% of damage in the most affected

Figure 6-1: San Francisco 1906, fire and earthquake damage.

SOURCE: KARL V. STEINBRUGGE COLLECTION AT NISEE. PHOTOGRAPHER: ARNOLD GENTHE.

areas of the city was caused by fire, with only 5% to 20% of the damage caused by shaking; the event was interpreted as a great fire rather than a great earthquake (Figure 6-1).

Differences in building performance based on construction, configuration, and soils conditions were observed during and after the earthquake, but there was no systematic investigation of the performance of anti-seismic systems voluntarily implemented by engineers in the last four decades of the 19th century, or of the code requirement implemented in 1901 to provide bond iron through the wythes of brick walls. Performance observations of selected steel-frame and concrete structures were used to justify the removal of some code restrictions concerning building height. In the aftermath of the San Francisco earthquake, engineers' awareness of seismic risk increased, resulting in voluntary efforts at seismic-resistant design, but codes did not specifically direct the design community to address earthquake related hazards.

Figure 6-2: Santa Barbara, 1925, typical failure of brick walls and timber interior.

SOURCE: KARL V. STEINBRUGGE COLLECTION AT NISEE; PHOTOGRAPHER: UNKNOWN.

## 6.2.2 The 1920s and the First Seismic Code

By the 1920s, the mechanisms for implementing seismic regulations for buildings in California were in place. The professional dialog about earthquake design had become more public since 1906, and post-earthquake investigators who examined field conditions after the 1925 Santa Barbara Earthquake called for code action (Figure 6-2).

In 1927, the Pacific Coast Building Officials Conference (precursor to ICBO, the International Conference of Building Officials) included an appendix of optional seismic design provisions in the first edition of the *Uniform Building Code* (UBC). A lateral load requirement was set at 7.5% of the building weight with an increase to 10% for sites with soft soils. This established the first version of the **equivalent lateral force procedure** still used in seismic codes today (Box 1). At the same time, some California cities began to adopt mandatory seismic design provisions in response to advocacy from citizens.

In the late 1920s the Structural Engineers Association of California (SEAOC) formed to address concerns about access to technical information and professional practice issues. For the next several decades, SEAOC's volunteer efforts would yield significant contributions to California codes and ultimately assume a leading role in the development of U.S. seismic codes.

## 6.2.3 Mid-Century Codes and the Introduction of Statewide Regulations

A large number of school buildings constructed of unreinforced masonry were severely damaged in the 1933 Long Beach earthquake (Figure 6-3). A public outcry for safer schools resulted in intense efforts by California legislators to quickly enact legislation requiring seismic design provisions. The Field Act of 1933 transferred the responsibility for approving plans and supervising construction of public schools to the State Division of

Figure 6-3: Long Beach, 1933, Alexander Hamilton Junior High School.

SOURCE: KARL V. STEINBRUGGE COLLECTION AT NISEE; PHOTOGRAPHER: UNKNOWN.

THE REGULATION OF SEISMIC DESIGN

## Box 1 The Equivalent Lateral Force Procedure

The Equivalent Lateral Force Procedure simplifies the dynamic effects of earthquakes by using a static model. Historically, the procedure was used for the design of all structures, but the current codes restrict its application to small buildings of regular configuration and larger buildings of limited height constructed with flexible diaphragms that are not considered to be essential or hazardous to the public.

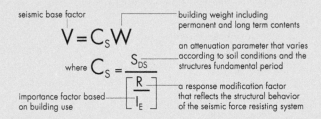

seismic base factor

building weight including permanent and long term contents

$$V = C_S W$$

an attenuation parameter that varies according to soil conditions and the structures fundamental period

$$\text{where } C_S = \frac{S_{DS}}{\left[\frac{R}{I_E}\right]}$$

importance factor based on building use

a response modification factor that reflects the structural behavior of the seismic force resisting system

In an earthquake, buildings experience ground motion that causes high accelerations and proportionately large internal forces in the building structure for short durations. In the equivalent lateral force procedure, static loads with a lesser magnitude than the actual earthquke forces are applied. This relies on the ability structures have to withstand larger forces for short periods of time and allows for a less conservative, more affordable seismic design. The seismic base shear V was specified as a given percentage of the building weight. The value is determined by combining factors representing properties of the structure, soil, and use of the building.

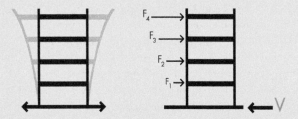

The tendency for the building to sway from side to side in response to ground motion produces greater accelerations in the upper parts of the building. This back and forth motion is called the fundamental mode, which dominates the response of most regular building structures. To model this effect statically, the equivalent lateral force procedure redistributes the load applied to the buildings floors to account for their distance from the base.

Architecture. It included lateral force requirements that varied according to the type of structural system. In the same year, the Riley Act created a mandatory seismic design coefficient for all buildings in the State of California and prevented construction of new unreinforced masonry buildings. The provisions of the Field and Riley Acts were developed in a simple form for implementation and did not reflect the latest developments in engineering understanding. It was not until 1943 that the Los Angeles Building Code adopted the first provisions in the United States that accounted for building height and flexibility.

The Long Beach earthquake provided the stimulus for state-mandated seismic design provisions that began the process of coalescing independent local efforts and assuring that minimum standards were enforced throughout California. However, the seismic provisions of the UBC did not become the standard in California until 1960. Before that time, many local jurisdictions added their own seismic design provisions to the Riley Act requirements.

In the late 1940s SEAOC responded to the inadequacy of seismic design codes by embarking upon work that would form the basis for the first edition of the *Recommended Lateral Force Requirements and Commentary*, also known as the SEAOC Blue Book. These recommended seismic design provisions for new buildings were included in the 1961 UBC. The Blue Book, published from 1959 through 1999, continued to evolve with major re-evaluations after significant earthquakes.

After 1960, the development of seismic design codes typically began with provision proposals initiated by the Blue Book that were later incorporated into the UBC. New systems, such as ductile moment frames, were incorporated into the code incrementally as they came into use. Inclusion of similar provisions in the other model codes in the United States followed the UBC's lead but occurred somewhat later. The UBC, used most extensively in the west, initiated national trends in earthquake-resistant code requirements.

The 1971 San Fernando earthquake occurred less than a decade after the extensively studied 1964 Alaska earthquake, and confirmed findings of greater-than-expected damage to engineered buildings meeting code provisions (Figure 6-4). Extensive data on the performance of newer building systems was collected. Observations stimulated research on the influence of reinforcement patterns on the strength and deformation of

Figure 6-4: Sylmar, 1971, Damage to engineered building at Olive View Medical Center.

SOURCE: KARL V. STEINBRUGGE COLLECTION AT NISEE; PHOTOGRAPHER: KARL V. STEINBRUGG.

concrete structural components. Evidence for the effects of building configuration and relative stiffness of structural elements as well as the performance of structural connections spurred revisions to code requirements.

Although the 1976 UBC increased building strength requirements and adopted the concept of increasing strength according to occupancy type, design forces remained only a fraction of the actual forces experienced by buildings during ground shaking (Figure 6-5). As discussed in chapter 4, building codes permit designs for less-than-expected earthquake forces to facilitate relatively simple linear analysis and design procedures because such designs, when coupled with detailing requirements also stipulated in codes, have proven successful in past earthquakes. In 1973, in an effort to assist the application of current technological develop-

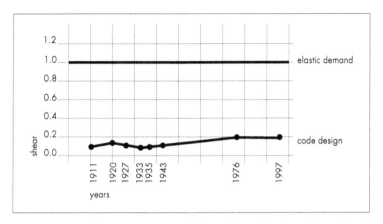

Figure 6-5: Graph comparing actual elastic demand with code-mandated design forces.

SOURCE: HOLMES, W., THE HISTORY OF U.S. SEISMIC CODE DEVELOPMENT, PUBLISHED IN THE *EERI GOLDEN ANNIVERSARY VOLUME 1948-1998*, EARTHQUAKE ENGINEERING RESEARCH INSTITUTE, OAKLAND, CALIFORNIA, 1998).

ments to structural engineering practice, SEAOC established the Applied Technology Council (ATC) to translate engineering research into usable design information.

In the aftermath of earthquakes of the 1960s and 70s, the need for more stringent building standards at the state level was recognized. The California Hospital Act, passed soon after the San Fernando earthquake, mandated more stringent building standards, plan checking, and inspection for hospitals under the direction of the Office of Statewide Health Planning and Development, with the intent of improving patient protection and maintaining building use through the regulation of nonstructural as well as structural design.

### 6.2.4 Late 20th Century: the Move toward New Model Building Codes

By the mid 1970s, the need for federal seismic design standards, coupled with the structural engineering profession's interest in significantly updating code content and streamlining its organization, shifted code development from regional to national efforts. A series of projects to develop national guidelines for seismic design began in the mid'70s. This effort provided the groundwork that led to the publication: ATC-3-

06, *Tentative Provisions for the Development of Seismic Regulations for Buildings.*
In 1977, the U.S. Congress passed the National Earthquake Hazards
Reduction Act. In the following year, FEMA established the National
Earthquake Hazard Reduction Program (NEHRP) and the Building
Seismic Safety Council (BSSC) to develop the first nationally applicable
seismic design guidelines for new buildings.

A major milestone was reached in 1985 when FEMA published the first
version of the NEHRP *Recommended Provisions for Seismic Regulations of
Buildings and Other Structures.* The *Provisions* were originally conceived
to be a resource document, rather than a code, but the format and
language of the *Provisions* conform to conventional code language. The
NEHRP Provisions had major national influence, and both the BOCA
(Building Officials Code Administrators) *National Code* and the SBCCI
(Southern Building Code Congress International) *Standard Code* adopted
requirements based on the NEHRP *Provisions* in 1991 and 1992 respectively. The seismic provisions of current building codes are largely based
on the NEHRP *Provisions,* supplemented by industry materials association standards. A significant difference between the NEHRP *Provisions*

Figure 6-6: San Francisco, 1989,
elevated highway support bents, an
example of failure of engineered
structures.

SOURCE: EERI ANNOTATED SLIDE COLLECTION

and earlier model codes was the introduction of provisions that related design forces to the characteristics of the ground motion of the site. This part of the *Provisions* required designers to consider dynamic effects and resulted in larger than previously considered forces and building deformations for some kinds of structures. Most important architecturally was the inclusion of regulations that identified building configuration as a factor in determining acceptable engineering analysis methods and selecting structural systems.

The extent of damage to buildings and the larger-than-expected economic losses caused by the 1989 Loma Prieta earthquake galvanized the need for public involvement in identifying acceptable building performance (Figure 6-6). After Loma Prieta, many building owners discovered that they misunderstood the protection provided by the codes and had made investment decisions based on unrealistic expectations about building performance and the likelihood of damaging earthquakes. A post-earthquake economic recession in Northern California alerted government officials to the potential economic cost of large earthquakes. The insurance industry responded with a lack of confidence in the insurability of the building stock. For those concerned with economic impacts, the life safety intent of building codes was no longer a sufficient standard. People who made decisions that affected their communities' seismic risk needed to better understand the relationship between code compliance and building performance. At the same time, engineers were developing methods for predicting building performance with greater accuracy. The combined effect of these developments stimulated an increased interest in performance-based codes and design guidelines.

Figure 6-7: Northridge, 1994 Failure of a welded steel connection.

The 1994 Northridge earthquake surprised the engineering community by severely damaging several recently constructed steel moment frame buildings (Figure 6-7). It confirmed the need for experimental research as part of the code development process. It also reinforced many of the lessons learned by the public in Loma Prieta. Damage patterns in Northridge revealed

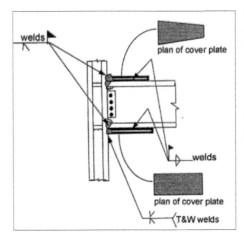

Figure 6-8: Example of a pre-qualified steel moment beam-column connection developed in the aftermath of the Northridge earthquake

SOURCE: SAC JOINT VENTURE, INTERIM GUIDLINES, REPORT # SAC -95-02 FEMA, WASHINGTON, DC.

the uncertainties associated with earthquakes, and provided examples of how buildings designed to resist the larger equivalent lateral forces required by more recent codes do not always perform better than older buildings. Ductile moment-resisting connections used in steel frames showed unexpected failures due to the complex behavior of welded beam-column connections.

In the aftermath of the Northridge earthquake, FEMA sponsored an extensive study of the steel moment connection problem. SAC, a joint venture of SEAOC, ATC and CUREE (Consortium of Universities for Research in Earthquake Engineering) conducted the work and produced a series of publications containing guidelines for engineering practice. The guidelines introduced pre-qualified connection designs and more explicit requirements for substantiating proposed designs with experimental research. The need for analytical techniques that could predict performance more accurately was identified. Many of the recommendations were later incorporated into building codes (Figure 6-8).

After the Loma Prieta and Northridge earthquakes, disaster-stricken communities raised questions about the nature of code protection, and design professionals began to question their own performance expectations for buildings designed to meet code requirements. Government agencies and representatives at the local, state, and national levels became concerned with the potential economic cost of large earthquakes. At the same time, innovative structural systems designed to improve performance, such as seismic isolation, became more commonly used.

In this climate, U.S. model code associations, the ICBO, the SBCCI , and BOCA, joined to form the International Code Council (ICC). By the year 2000, the UBC and NEHRP seismic design provisions were merged in the first edition of the *International Building Code* (IBC). Shortly after, NFPA (National Fire Protection Association) undertook a building code development process that produced the *NFPA 5000 Building Code* with seismic design requirements also based on the NEHRP *Provisions*. As local jurisdictions adopt these new model codes design teams, working on projects in regions with lower levels of seismicity, have been required to address more stringent seismic design regulations than has been the case in the past, particularly on sites with less than optimal soils conditions. NEHRP requires site-specific soil data in order to determine the seismic design category of a building. According to NEHRP, a building project in Atlanta on soft soil may be required to have the same level of seismic design as a building in California. More extensive geotechnical investigations of the building site can be used to determine the appropriate code-defined soil classification and provide the design team with critical information that will affect seismic design requirements specified by the code.

In the 1990s, FEMA sponsored the development of guidelines for the seismic evaluation and rehabilitation of existing buildings that introduced methods that would inform the future conceptual basis of codes for new buildings. Nonlinear analysis, an analytical method that integrates the deformation of a structure into the analysis of a structural design, was identified as an essential tool for some seismic design applications and the concept of performance goals was introduced. This work produced several publications that have had a significant impact on code development: FEMA 273, NEHRP *Guidelines of the Seismic Rehabilitation of Buildings*, supplemented by FEMA 274, a Commentary, and FEMA 276, Example Applications. FEMA 356, *Prestandard and Commentary for the Seismic Rehabilitation of Buildings* converted the guidelines into code language. They are discussed in Chapter 8. Many of FEMA's seismic evaluation and rehabilitation concepts have been adopted in the *International Existing Building Code.*

By the end of the decade, engineering organizations put forward several proposals for code action that called for a more unified, performance-based approach to the regulation of seismic design for new buildings. (See section 6.4 for a discussion of performance-based seismic design.) The SEAOC *Vision 2000: Performance-Based Seismic Engineering of Buildings* called for a code development project that would create a more coherent rationale for seismic code provisions, drawing upon new analytical methods and incorporating an approach that incorporates performance predictions into the design process. The Earthquake Engineering Research Center (EERC) summarized the issues surrounding building performance in FEMA 283, *Performance-Based Seismic Design of Buildings: an Action Plan for Future Studies.* The Earthquake Engineering Research Institute (EERI) produced FEMA 349, *Action Plan for Performance-Based Seismic Design*, and FEMA has now contracted for the development of performance-based seismic design guidelines.

## 6.2.5 Current Status of Seismic Code Development

The introduction of *IBC* and *NFPA 5000* has assisted the move toward standardizing seismic code regulations in the U.S., but design professionals express concern that the pace of code developments has not kept up with the profession's needs. Debates about the value of simplicity versus the value of reliability continue. Critics argue that codes have become unnecessarily complex and fail to communicate the intent behind code provisions. There is concern that code procedures can be overly restrictive in ways that discourage sound design strategies. There is also concern that, in some cases, designs that "meet the code" may be inadequate. These issues are being aired and addressed in the dialog that surrounds the code revision process and in the work currently underway to develop methods of performance-based design.

## 6.3 CODE INTENT

### 6.3.1 The Purpose of Earthquake Code Provisions

The IBC's stated purpose is:

> "to establish the minimum requirements to safeguard the public health, safety and general welfare through structural strength, means of egress, facilities, stability, sanitation, adequate light and ventilation, energy conservation and safety to life and property from fire and other hazards attributed to the built environment and to provide safety to fire fighters and emergency responders during emergency operations. "

The primary intent of all seismic code provisions is to protect the life safety of building occupants and the general public through the prevention of structural collapse and nonstructural life-threatening hazards during an earthquake. However, it is generally acknowledged that seismic code provisions are also intended to control the severity of damage in small or moderate earthquakes. In large earthquakes, damage is expected; engineers rely on the mechanisms provided by damage to contribute to a structure's damping capacity. The SEAOC *Blue Book* states that code-designed buildings should be able to:

1.  Resist a minor level of earthquake ground motion without damage;

2.  Resist a moderate level of earthquake ground motion without structural damage, but possibly experience some nonstructural damage;

3.  Resist a major level of earthquake ground motion having an intensity equal to the strongest either experienced or forecast for the building site, without collapse, but possibly with some structural as well as nonstructural damage.

Codes also recognize that some structures are more important to protect than others. The NEHRP *Provisions* stated purpose is:

1.  To provide minimum design criteria for structures appropriate to their primary function and use, considering the need to protect the health, safety, and welfare of the general public by minimizing the earthquake-related risk to life, and

2.  To improve the capability of essential facilities and structures containing substantial quantities of hazardous materials to function during and after design earthquakes.

### 6.3.2 Conflicts Between Intent, Expectations, and Performance

Codes do not explicitly address economic intent. Members of the general public who believe earthquake-resistant design should provide them with usable buildings after an earthquake often misunderstand the term "meets code". Design professionals are responsible to convey seismic design performance expectations to decision makers. Clear communication between engineers and building owners is important and the architect's role as a facilitator of the dialog between owners, and members of the design team is critical to promoting a shared understanding that can form the basis for appropriate design decisions. This communication is complicated by the indirect and somewhat unpredictable relationship between code compliance and building performance. The NEHRP *Provisions* state:

> "The actual ability to accomplish these goals depends upon
> a number of factors including the structural framing type,
> configuration, materials, and as-built details of construction."

Local site factors and the variations that can occur with different combinations of structural systems and materials, construction methods and building configuration cause differences in building performance. Factors influencing performance can be subtle and the cause of damage difficult to determine. Earthquakes have produced different damage patterns in apparently identical adjacent buildings. The professional judgment of the design team, the extent of code compliance, and level of plan review also affect building damage. For example, school buildings constructed under the quality control regulations of the Field Act have consistently outperformed other buildings designed to meet similar code provisions.

Given the complexity of building performance in earthquakes, it is readily apparent that many past code "improvements" which increased the lateral base shear (and therefore the strength of buildings to resist a static lateral load) did not have a directly proportional effect on building performance. There are too many other factors involved. Although more sophisticated analysis methods can produce better performance predictions, seismic performance of buildings can be more accurately expressed as a probabilistic rather than as an absolute phenomenon. For example, in a Modified Mercalli level IX earthquake (defined in Chapter 2), it is expected that less than one percent of the midrise concrete shear-wall buildings designed to meet the seismic code minimums of the 1991 UBC will collapse. Five percent of them will experience extensive structural and nonstructural damage, and thirty percent will experience moderate (primarily nonstructural) damage. Seventy percent will have minor or no damage. At present, performance data from real earthquakes or analysis results is not extensive enough to generate reliable probabilistic scenarios for all of the existing and commonly designed combinations of building configurations and structural systems. However, as this information becomes available, it is likely to be reflected in future codes.

Codes are put forward as minimum standards, but it is common practice for designers and owners to aim to meet rather than exceed code minimums as way to control project costs. As owners have become more aware of the limitations of seismic codes, there is greater interest in electing to build to higher standards. This is particularly true for corporate and institutional building owners who self-insure or include risk management and loss projections in their planning process. As owners raise concerns about acceptable levels of risk, there is an increased need to be able to predict building performance and relate that performance to design standards.

## 6.4 PERFORMANCE BASED SEISMIC DESIGN

### 6.4.1 Prescriptive Design, Performance Design, and the Code

Performance-based design is fundamentally different from prescriptive design. Prescriptive codes describe what to do—the goal is to achieve a particular design outcome that meets the intent of the code. For example, Table 2305.3.3 of the IBC states that the maximum aspect ratio (height to length) for shear walls sheathed with particleboard is 3.5:1. The intent behind this provision is to insure adequate sheathing of elements used as shear walls and to prevent unrealistically high tie-downs from overturning moments. In contrast, a performance approach for the same structure would describe the intent of the code in a way that allows the designer to decide how the intent is met. In the case of the wood frame building above, the code might stipulate that the lateral movement, or drift, of each floor with respect to the floor below may not exceed 2.5% of the floor-to-floor height. It is then up to the designer to decide how to achieve this outcome. If the designer chooses to use shear walls that are more slender than prescribed by the code, then the designer will need to demonstrate, using a rational basis acceptable to the building official, that the slender shear walls, as well as the rest of the structure, will meet performance requirements.

The distinction between performance and prescriptive methods is complicated by the fact that performance can be a relative concept. For example, a designer studying a concrete-frame building with shear walls could view the maximum story drift permitted by codes for this structural system as a prescriptive requirement. The reason the code limits drift is to prevent the negative consequences of large lateral movements. These include p-delta effects that compromise column integrity, and the concentrated stresses at connections, and higher levels of nonstructural damage caused by excessive movements of the structural frame. If a designer can address these issues through a performance-based approach, the drift need not be limited.

Prescriptive code provisions have some advantages. They have been shown to be reliable for meeting life safety objectives in the U.S. The performance of school buildings in recent California earthquakes substantiates this. They can be applied consistently even in cases where design judgment is difficult or where the designer or building official is inexperienced with alternative design methods. But prescriptive codes do not readily support innovative or alternative approaches to seismic design that provide equal or better building performance.

## 6.4.2 Definitions of Performance-Based Seismic Design

The term "performance based seismic design", as currently used, has multiple definitions. It is used to refer to a design approach that meets the life safety and building performance intents of the code, while providing designers and building officials with a more systematic way to get at the alternative design option currently available in codes. In this regard, performance-based seismic design facilitates innovation and makes it easier for designers to propose new building systems not covered by existing code provisions or to extend the use of existing systems beyond code limitations. For example, a designer may propose to use a given structural system for a building that is taller than code permits for that system. The designer would provide the building official with a performance-based rationale that shows how the design will meet the intent behind the code's height limit.

Performance-based design is also used to refer to a design approach that identifies and selects a performance level from several performance options. The current version of the IBC can be called performance-based because it incorporates distinctions between performance goals for different building uses. But the term is more commonly used when referring to performance options that exceed code minimums or in cases when buildings are expected to remain operational after a disaster.

From a technical perspective, performance-based design has a different definition. It is a design approach that provides designers with tools to achieve specific performance objectives such that the probable performance of a structure could be reliably predicted. Current codes do not aim to do this. Their goal is to achieve a minimum standard, based on life safety, for most structures and an immediate post-earthquake occupancy standard for facilities essential for post-earthquake recovery. Code requirements have evolved over time to insure that a reasonable effort has been made to meet the minimum standard, but they do not yield performance predictions. The expected performance level of a building that meets current seismic codes is highly variable and undefined by the code. However, the seismic engineering community is now exploring code development options that create a more explicit link between design approach and code content.

### 6.4.3 Implementing Performance-Based Seismic Design

At the present time, code development efforts related to seismic design are focusing on incorporating performance-based design concepts. It is proposed that future codes would establish frameworks that would assist designers as they provide building owners with a clear choice between a minimum standard and specified higher performance levels. This work includes an effort to translate recent advances in engineering understanding of building performance in past earthquakes, laboratory tests, and structural analysis methods to design guidelines. FEMA, seeking to improve code reliability and facilitate design to higher standards, is funding a longer term effort to develop performance-based design guidelines as initially outlined in FEMA 349. The project scope includes the development of structural, nonstructural, and risk management guidelines supported by a development plan and a stakeholder's guide.

The shift to a performance-based seismic design code will have other broader impacts on building design practice. Although architects are generally not concerned with the details of applying the seismic code and routinely delegate this aspect of code compliance to the structural engineer, a performance-based seismic code will involve the architect more directly in the seismic design process. Architects working with such a code must become very familiar with the definition of performance levels, and the economic implications of the choice of level. Seismic design concerns are also likely to figure more explicitly in the pre-design phase of the architectural design process when project objectives are identified, which often takes place before the involvement of the structural engineer. In the traditional role of coordinator of the design team and primary contact for the client, the architect will facilitate the dialog surrounding performance-based seismic design decisions. As the primary deliverer of information to the client, the architect will need to include more extensive discussions of technical and economic aspects of seismic design than are presently necessary with a single standard code. In projects with alternative administrative structures, such as design-build, architects will experience increased interaction related to seismic design decisions.

As performance-based seismic design methods come into use, the technical challenges inherent in devising structures for architectural schemes that require structural irregularities may increase. As discussed in Chapter 5, buildings with regular configurations perform more predictably than those with irregular configurations. Engineers attempting to meet performance levels specified in a code may become more reluctant to provide performance assurances for irregular structures proposed by the architect. The pressure to predict performance accurately could result in an overly conservative approach to building configuration. Communication and creative collaboration between the architect and the engineer will become more critical, particularly during the initial conceptual phases of the building design process.

The shift to a performance-based approach will also impact the format, enforcement, and implementation of codes. Current codes do not specify design methods or imply a design philosophy. They presume that designers will determine the design approach. A performance-based code suggests that the criteria associated with performance objectives would emerge from an articulated conceptual framework for seismic design. The format of codes would reflect that framework. Enforcing a code that specifies multiple levels of design and performance would require a more complex procedure than currently used. Implementation of a performance-based seismic design code suggests a departure from the traditional evolutionary model of code development to a redesign of the code document. This process would require the support of elected officials and the public. Its success will depend upon increased public understanding of the behavior of structures in earthquakes, the limitations of current codes, and the rationale behind a performance-based approach.

A performance-based seismic design code format could provide a unified basis for comparison of design alternatives that give decision makers a consistent means of quantifying risk. That basis will enable design professionals to respond to the needs of all decision makers and stakeholders concerned with seismic design, including owners, lenders, insurers, tenants, and communities at large. Performance-based design methods are already being used for this purpose, but a code-specified framework could assist decision makers who wish to compare design alternatives and project types.

The need for simplified methods for lower risk building project types will remain, and performance-based seismic design would ultimately provide a basis for the development of new prescriptive code requirements. These requirements would be as simple or simpler to apply as current code provisions, but would have a unified conceptual basis that would provide more predictable performance outcomes. It would be easier for designers to relate code provisions to code intent and eliminate some of the obstacles to seismic design innovation. It would also allow designers to combine performance-based and prescriptive methods in a more consistent manner.

## 6.5 SEISMIC DESIGN PROVISIONS

### 6.5.1 Code-Defined Parameters

Seismic design provisions vary in complexity, depending upon the risk associated with the project type. As building risk increases and structural design demands become greater, code provisions expand to include a greater number of parameters that support more sophisticated analytical techniques. Table 6.1 illustrates how these code-defined parameters can be influenced by architectural design decisions, thereby impacting seismic design requirements.

Table 6-1: The effects of architectural design decisions on code parameters and code-mandated design requirements

| IBC SEISMIC DESIGN PARAMETERS | ARCHITECTURAL DESIGN DECISIONS | SEISMIC DESIGN REQUIREMENTS |
|---|---|---|
| Ground motion accelerations | Site selection at the national and regional scale | Affects design earthquake forces |
| Site classes (soils properties) combined with ground motion accelerations to determine a site coefficient | -Site selection at regional and local levels<br>-Placement of the building on a particular site | -Failure-prone soils require site-specific geotechnical investigation<br>-Site coefficient affects design earthquake forces<br>-Soils properties affect building response to ground motion |
| Fundamental period of the structure | -Building height<br>-Selection of structural system | -Affects design earthquake forces<br>-Affects building response to ground motion |
| Seismic use groups and occupancy importance factors | Assignment of program spaces to buildings | -Affects eligibility for simplified analysis methods<br>- Triggers need to meet more stringent code requirements |
| Seismic design category that relates structure importance to design accelerations | Site selection for particular building uses | Used to identify appropriate code procedures |
| Building configuration classification | -Building size<br>-Footprint geometry and massing<br>-Organization of interior spaces<br>-Structural framing patterns | Used to modify the analysis procedures specified by the code and can trigger more extensive analysis requirements |
| Response modification factor, system over strength factor, deflection amplification factor, redundancy coefficient | -Type of lateral load resisting system<br>-Materials of construction of the lateral load resisting system | -Affects design earthquake forces<br>-Affects building response to earthquake forces |

## 6.5.2 Performance Levels

In performance-based seismic design, a decision-making team that includes designers as well as stakeholders makes choices between alternate performance levels. To do this, the team must consider the appropriate level of seismic hazard to which the building should be designed as well as the acceptable risk that would guide building performance expectations. Implicit in this process is an evaluation of costs and benefits. Performance levels must be articulated qualitatively and technically. Qualitative performance objectives are stated according to the needs of stakeholders. These objectives include life safety, ability to use a building for shelter after an event, ability to continue to produce services or income at the site after an event, and the costs associated with repair, loss of use, and loss of contents.

Technical performance levels translate qualitative performance levels into damage states expected for structural and nonstructural systems. As defined in Table 6-2, the SEAOC *Vision 2000* document proposes four qualitative performance levels. Figure 6-9 shows how the NEHRP *Provisions* relate these performance levels to probabilistic seismic hazard levels and occupancy categories called Seismic Use Groups (SUG). SUGs classify structures according to risk and importance. SUG I is the category

Table 6-2: SEAOC *Vision 2000* Qualitative Performance Levels

| Fully operational | Continuous service. Negligible structural and nonstructural damage. |
|---|---|
| Operational | Most operations and functions can resume immediately. Structure safe for occupancy. Essential operations protected, non-essential operations disrupted. Repair required to restore some non-essential services. Damage is light. |
| Life Safety | Damage is moderate, but structure remains stable. Selected building systems, features, or contents may be protected from damage. Life safety is generally protected. Building may be evacuated following earthquake. Repair possible, but may be economically impractical. |
| Near Collapse | Damage severe, but structural collapse prevented. Nonstructural elements may fall. Repair generally not possible. |

SOURCE: SEAOC VISION 2000 REPORT

THE REGULATION OF SEISMIC DESIGN

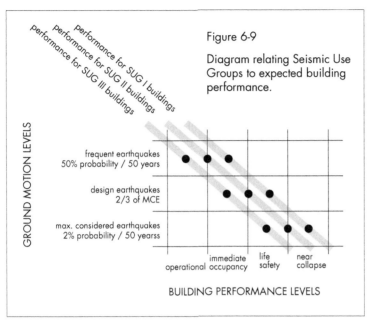

Figure 6-9

Diagram relating Seismic Use Groups to expected building performance.

SOURCE: FEMA 450 PART 2, *COMMENTARY ON THE 2003 NEHRP RECOMMENDED PROVISIONS FOR NEW BUILDINGS AND OTHER STRUCTURES,*

assigned to most structures. SUG II structures have high-occupancy levels with restricted egress. SUG III structures are essential facilities for post-earthquake recovery or facilities that contain hazardous materials. The performance criteria for each group is progressively more stringent.

## 6.5.3 Performance-Based Seismic Engineering

The technical definition of a performance level must specify building performance in terms that can be verified by an analysis of the proposed building design. As the design team responds to the specific needs of individual stakeholders, multiple performance objectives will be identified, and multiple technical parameters will be used to establish acceptance criteria. The identification and implementation of technical performance criteria requires the ability to predict, with reasonable reliability, the behavior of a building subject to an earthquake hazard. Recent advances in analytical tools and computational capabilities have advanced the art of earthquake performance prediction and made the development of performance-based approaches to engineering design more possible now than in the recent past.

Seismic engineering analysis is the art of translating understandings of seismic hazards into predictions of building behavior in a way that can inform building design. It is used in the conceptual design phase to evaluate alternative structural systems and configurations, in the schematic design phase to refine the layout of the selected structural system, and in the design development phase to determine the details of construction. Throughout the design process, engineering analysis is used to verify that the proposed design will perform at an acceptable level and satisfy building code requirements. Engineers select analysis methods based on the type of structural behavior being examined and the need for information about that behavior. Simple methods can be useful for conceptual design even when complex methods are required for design development. In a given building design project, multiple analysis methods will be used to verify design decisions that range from the selection of a structural system to the selection of a bolt diameter.

In performance-based seismic engineering practice, performance implications of some design decisions have to be predicted with a greater degree of accuracy than others. Designers aim to use the simplest, most cost-effective method appropriate for the design task at hand. It takes a combination of theoretical knowledge and engineering judgment to select appropriate analytical methods. Routine design of small regular buildings of conventional construction can be accomplished with simple analytical methods, whereas large, irregular buildings are likely to require a more complex analysis. Designers will sometimes choose to use more than one analysis method as a way to examine different aspects of a structure's behavior. Although the code does not specify approaches to seismic design, it does restrict the use of certain analytical methods to insure that they are used appropriately. From an architect's perspective, this means that some combinations of building parameters will require more costly engineering analysis methods.

## 6.5.4 Engineering Analysis Methods

Seismic analysis methods can be divided into two groups—linear procedures and nonlinear procedures. Linear procedures are by far the most common and are used in the majority of seismic design applications for new buildings. The term "linear" refers to the assumption that there is a constant proportional relationship between structural deformation and the forces causing it. Engineers are familiar with linear analysis because it is the same as used for analysis of beams, girders, and columns for gravity systems. Most prefer linear seismic design for the same reason. However,

## Box 2. Linear Dynamic Analysis

A linear dynamic analysis is useful for evaluating irregular or dynamically complex buildings. An irregular building is defined as having a distribution of mass or stiffness of the structure that is nonuniform and is often created in buildings that that have complex space planning requirements or asymmetrical configurations. Dynamic complexity is common in flexible structural systems. Flexibility is greatly influenced by the selection of structural system and building height. Flexible buildings tend to have a significant response to higher mode shapes. Mode shapes are movement patterns that occur naturally in structures that have been set in motion by ground shaking. The diagram below compares the shapes of the first three modes.

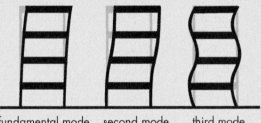

first or fundamental mode     second mode     third mode

Designers use linear dynamic analysis to determine the degree of influence each mode shape will have on a structure's performance. The importance of higher modes depends on the relationship between the fundamental mode of the structure and the dynamic-ground shaking characteristics of the site. Designers express mode shape influence in terms of the percent of building mass assigned to a particular mode. If the building mass vibrates primarily in the first or fundamental mode, a static analysis is permitted by the code. Although linear dynamic analysis methods are becoming routine in engineering practice, they are more complicated because they require detailed information about ground motion. When linear dynamic analysis is used to meet code requirements or check code conformance, the structure-ground shaking interaction is usually modeled using a response spectrum. The IBC code includes design response spectra, a procedure for developing a design structure curve based on the general design response spectrum show below.

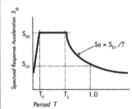

design response spectrum

An alternative and significantly more complex method for modeling ground shaking, called a time history analysis, examines modal response using actual ground motion data. The code requires that time history analyses consider several different ground motion records to insure that the structure response is sufficiently representative to account for future unknown ground motion patterns.

---

the linear forces used and building deformations are not those expected in an earthquake and have been somewhat arbitrarily developed to result in buildings that are adequately resistant to life-threatening earthquake damage. A nonlinear approach adds another layer of analytic complexity. It takes into account the changing stiffness of various elements and the overall structure during the shaking. Structural elements are designed by examining deformations rather than forces. Although nonlinear analysis can be used to design new buildings, it is more commonly used to predict seismic response for evaluation and retrofit of existing structures.

Linear analysis can be static or dynamic. The IBC's equivalent lateral force procedure, shown in Box 1, is a simple version of a linear static analysis. A linear static procedure is well suited to buildings with regular configurations that have a response to ground motion that is dominated by the back and forth swaying of a structure called the fundamental mode. Linear models can also be dynamic. The dynamic version of a linear model, shown in Box 2, is known as a modal analysis. It is based on an idealized site response spectrum and takes into account motions that are influenced by higher mode shapes providing more information about a building's behavior under seismic loads. Because it is linear, the displacements expected under different modes can be added together to identify critical design behaviors. A modal analysis is frequently used to create a more accurate picture of how irregular structures would perform. Seismic codes allow and, in some cases, require designers to substitute a modal analysis for static methods.

Nonlinear analysis methods can also be static or dynamic. The "push-over" analysis shown in Box 3 is a static nonlinear analysis technique that offers the potential for increased accuracy at the expense of increased complexity. It requires multiple iterations of a static analysis that can account for the effects of yielding elements and help designers visualize how building behavior transforms as damage states progress. It is often used to identify and control the weak links that initiate structural failure mechanisms.

A nonlinear dynamic analysis or time history analysis is the most sophisticated and time-consuming method requiring a detailed knowledge of building properties and ground motions. It is relatively new to design practice and is used in the research and experimental stages of design applications. The IBC accepts this analysis procedure, provided that a design review is performed by an independent team of qualified design professionals.

THE REGULATION OF SEISMIC DESIGN

## Box 3. Nonlinear Static "Pushover" Analysis

A pushover analysis is a nonlinear static method that accounts for the way structures redesign themselves during earthquakes. As individual components of a structure yield or fail, the earthquake forces on the building are shifted to other components. A pushover analysis simulates this phenomenon by applying loads until the weak link in the structure is found and then revising the model to incorporate the changes in the structure caused by the weak link. A second iteration is then performed to study how the loads are redistributed in the structure. The structure is "pushed" again until the second weak link is discovered. This process continues until the yield pattern for the whole structure under seismic loading is identified.

A pushover analysis is only useful for evaluating nonlinear structures for which the fundamental mode dominates and is not suitable for certain irregular or dynamically complex structures. These would need to be designed using a nonlinear dynamic method. Some building structures are inherently linear because they lack redundancy or exhibit brittle modes of failure. These structures can be modeled sufficiently accurately using linear analysis methods. A sequence of nonlinear events used in a pushover analysis is shown below for a single-story, two-bay reinforced concrete frame with one bay of steel cross bracing. The sequence is shown numbered in the figure.

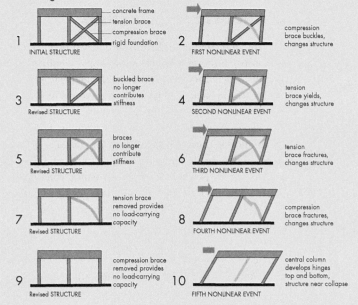

In current engineering practice, pushover analysis is more commonly used to evaluate the seismic capacity of existing structures and appears in several recent guidelines for seismic design concerning existing buildings. It can also be a useful design tool for the performance-based design of new buildings that rely on ductility or redundancies to resist earthquake forces.

To implement this method reliably, designers must be able to make a sufficiently accurate model that will reflect inelastic properties and potential deformation mechanisms of the structure. This requires a more sophisticated understanding of the structure's behavior than linear methods. Nonlinear analysis results produce more data and can be difficult to interpret. The current IBC code treats nonlinear analysis as an alternate method, but nonlinear approaches are likely to become integrated into the code as the code evolves in the future.

ADAPTED FROM A PRESENTATION MADE BY RON HAMBURGER AT EARTHQUAKE ANALYSIS METHODS: PREDICTING BUILDING BEHAVIOR, A FEMA-SPONSORED EERI TECHNICAL SEMINAR, 1999.

In general, nonlinear methods are not useful if the intent of the design is to verify compliance with minimum code provisions. Nonlinear methods generate a significantly greater quantity of analytical results that require substantially more engineering effort to produce and interpret. These methods are more appropriately used to determine estimates of performance with higher reliability than standard code expectation or to analyze the combined effects of new retrofit components and existing systems. If the intent of the analysis is to verify code compliance, linear analysis methods will suffice. Future codes are likely to include explicit guidelines for the use of nonlinear methods as they become a more routine part of conventional engineering practice.

## 6.6 NONSTRUCTURAL CODES

The IBC 2003 seismic code deals with the problem of reducing damage to nonstructural components in two ways. The first is to impose limits on the horizontal drift or deflection of the main structure. This is because nonstructural damage (such as glass breakage or fracturing of piping) may occur because the building structure is too flexible, causing racking in wall panels, partitions, and glazing framing. Flexible structures are economical because the code allows for considerably reduced forces to be used in the design; however, this approach solves the structural problem at the expense of the nonstructural components. Although this approach can still be used, the imposition of drift limits ensures that the flexibility of the structure will not be such that excessive nonstructural damage results.

The second approach is to assign force values based on acceleration to the critical nonstructural components and their connections to ensure that they will be strong enough to resist seismic accelerations in their own right or as the result of attachment to the structure. The analysis procedure is similar to, although less complex than, the procedure for determining the equivalent lateral force on the main structure. Modifications to the basic F=Ma equation include coefficients for importance, component amplification, component response modification, and the height of the component within the structure. All these modifiers increase the design forces relative to the spectral accelerations derived from the hazard maps.

In a way similar to that used to determine the forces on the main structure, the required performance characteristics for nonstructural components are related to three different Seismic Use Groups based on the use and occupancy of the building.

Several exemptions are made in the IBC seismic code:

1. All components in Seismic Design Category A structures are exempt because of the lower seismic effects on these items.

2. All architectural components (except parapets supported by bearing or shear walls) in Seismic Design Category B structures are exempt if they have an importance factor of 1.00, which indicates that they are not a life safety threat. (The importance factor is a 1.5 multiplier for components that are needed to function after the earthquake for life safety, that contain hazardous contents, or are large storage racks open to the public.). The importance factor is selected by the engineer, based on criteri in the code and consultation with the building official.

3. Mechanical and electrical components, Seismic Design Category B.

4. Mechanical and electrical components, Seismic Design Category C when the components importance factor is 1.00.

5. All Mechanical and electrical components in all seismic design categories that weigh less than 400 pounds, are mounted 4 ft. or less above the floor, have an importance factor of 1.00, are not critical to the continued operation of the building, and include flexible connections between the components and associated ductwork, piping and conduit.

6. Mechanical and electrical components in Seismic Design Categories C, D and F that weigh 20 pounds or less, the importance factor is 1.00, and flexible connections are provided.

Of all the elements of the building envelope, heavy precast concrete wall cladding panels attached to steel or reinforced concrete-frame structures require the most design and construction attention to ensure seismic safety. These typically span from floor-to-floor: horizontal drift or deformation of the building structural frame can create considerable racking forces in panels that are rigidly attached at top and bottom, resulting in damage or possible drop-off. Therefore, the attachment of these panels must permit differential movement of the floors without transmitting racking forces to the panels. This is achieved by special detailing of the connection of panels to structure.

Seismic codes require that heavy panels accommodate movement either by sliding or ductile connections. In high seismic zones sliding connections are rarely used, because of the possibility of incorrect adjustments when bolts are used, jamming or binding due to unwanted materials left after installation, and jamming due to geometrical change of the structural frame under horizontal forces.

The need for disassociating the heavy panel from the frame has a major impact on connection detailing. As a result, a connection commonly termed "push-pull" has been developed, primarily in California, which provides, if properly engineered and installed, a simple and reliable method of de-coupling the panel from the structure. The generic connection method consists of supporting the panel by fixed bearing connections to a structural element at one floor to accommodate the gravity loads, and using ductile "tie-back" connections to a structural element at an adjoining floor (Figure 6-10).

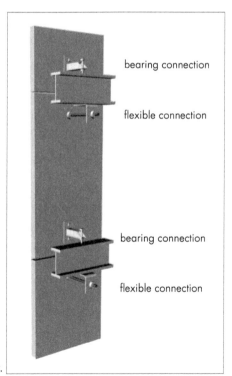

bearing connection

flexible connection

bearing connection

flexible connection

Figure 6-10

Typical floor-to-floor push-pull panel connections. Each beam has a bearing connection at the bottom of a panel and a flexible, or tie-back, connection for the panel below.

Recent developments in nonstructural seismic codes include a performance-based design approach, comparable to that used for structural design. This approach was first defined in the NEHRP *Guidelines for the Seismic Rehabilitation of Buildings* (FEMA 273,274) and subsequently in FEMA 356, *Prestandard and Commentary for the Seismic Rehabilitation of Buildings*. This will be replaced in 2006 by a new ASCE standard, ASCE 41.

## 6.7 CONCLUSION

The continually transforming content of seismic design codes reflects the evolution of design practice as it takes place in changing technical and political contexts. As the shift to a performance-based approach takes place, designers are raising questions about possible impacts on design practice. How will a performance-based seismic code affect professional liability? How will it affect the cost of professional design services? Will it be particularly difficult or expensive for small firms or inexperienced owners to implement? Will it really help designers manage uncertainty? Without real data on real buildings in real earthquakes to confirm our performance predictions, how confident can we be that our performance-based design methods work?

As the dialog surrounding building code development continues, new insights are emerging, particularly the recognition that this process needs the attention of all stakeholders concerned with the built environment. For architects, the new codes and the performance-based concepts behind them will require greater involvement in seismic design decisions. As architects help owners investigate the feasibility of proposed building projects and lead their clients through the design process, they will need to be aware of the interaction between design decisions and seismic design regulations.

## 6.8 REFERENCES

ACSE, 2003. *Minimum Design Loads for Buildings and Other Structures*, prepared and published by the American Society of Civil Engineers, Reston, Virginia.

ASCE, 2000. *Prestandard and Commentary for the Seismic Rehabilitation of Buildings*, prepared by the American Society of Civil Engineers, published by the Federal Emergency Management Agency, FEMA 356, Washington, D.C.

ATC, *Communicating with Owners and Managers of New Buildings on Earthquake Risk*, prepared by the Applied Technology Council for the Federal Emergency Management Agency, FEMA 389, Washington, D.C.

ATC/BSSC, 2003, *NEHRP Guidelines for the Seismic Rehabilitation of Buildings*, prepared by the Applied Technology Council for the Building Seismic Safety Council, published by the Federal Emergency Management Agency, FEMA 273, Washington, D.C.

ATC/BSSC, 1997. *NEHRP Commentary on the Guidelines for the Seismic Rehabilitation of Buildings*, prepared by the Applied Technology Council for the Building Seismic Safety Council, published by the Federal Emergency Management Agency, FEMA 274, Washington, D.C.

BOCA, *National Building Code*, published by the Building Officials and Code Administrators, Country Hills, Illinois.

BSSC, 2004. *NEHRP Recommended Provisions for Seismic Regulations for New Buildings and Other Structures, Part I: Provisions, and Part II, Commentary*, 2000 Edition, prepared by the Building Seismic Safety Council, published by the Federal Emergency Management Agency, Publications, FEMA 450, Washington, DC.

EERC, 1996. *Performance-Based Seismic Design of Buildings: An Action Plan for Future Studies*, prepared by the Earthquake Engineering Research Center and published by the Federal Emergency Management Agency, FEMA 283, Washington D.C.

EERI, 2000. *Action Plan for Performance Based Seismic Design*, prepared by the Earthquake Engineering Research Institute and published by the Federal Emergency Management Agency, FEMA 349, Washington D.C.

ICBO, 1997. *Uniform Building Code,*1997 edition prepared and published by the International Council of Building Officials, Whittier, California.

ICC, *2002, 2003. International Building Code*, prepared and published by the International Code Council, Falls Church, Virginia.

IEBC, 2003. International Existing Builing Code, published by the International Code Council, Falls Church, Virginia.

*NPFA, 2002. NFPA 5000 Building Construction and Safety Code*, prepared and published by the National Fire Protection Agency, Quincy, Massachusetts.

SBCCI, *Standard Building Code*, published by the Southern Building Code Congress International, Birmingham, Alabama.

SEAOC, 1999. *Recommended Lateral Force Requirements and Commentary*, prepared by the Structural Engineers Association of California, published by the International Conference of Building Officials, Whittier, California.

SEAOC, 1995.*Vision 2000: Performance Based Seismic Engineering of Buildings*, Structural Engineers Association of California, Sacramento, California.

by Eric Elsesser

## 7.1 INTRODUCTION

Design of any building is a challenge for architects and engineers, and the challenge is made more complex by providing for earthquake resistance. During the past 100 years, seismic design philosophy and details have progressed from simply considering earthquakes to be the same as wind loads, to a sophisticated understanding of the phenomenon of the earthshaking that induces a building response.

This chapter covers the 100-year history of seismic structural systems as developed by engineers and architects, ranging from simple to sophisticated solutions. Basic structural behavior is outlined; guidance for selecting a good structural system is suggested, and the following issues are discussed:

○ Scale and size of buildings and structural components

○ The impact of building configuration

○ Force verses energy

○ Drift or movement

○ Structural mechanisms (passive to active)

○ Costs and post-earthquake repair costs

## 7.2 A BRIEF SUMMARY OF 100 YEARS OF STRUCTURAL SEISMIC DESIGN

Seismic exposure has extended over many centuries, but systematic seismic design has occurred only over the past 100 years, especially in California since the 1906 San Francisco earthquake.

A group of thoughtful and creative engineers, responding to the observed damage in the 1906 earthquake, started to study, conceive, and design a progression of structural solutions to solve the earthquake response problem. This creative work has extended over a 100-year period, and continues today. A brief progression of key milestones in this seismic design history follows:

○ Initial seismic designs for buildings were based on wind loads, using static force concepts. This approach started in the late 1800s and lasted to the mid-1900s.

○ After the 1906 San Francisco earthquake, concepts of building dynamic response gained interest, and in the early 1930s, initial studies of structural dynamics with analysis and models were initiated at Stanford University. This approach ultimately led to a design approach that acknowledged the importance of building periods and dynamic rather than static design concepts.

○ Dynamic design concepts were enhanced by the acceleration spectra method used for design as developed by Professor Housner at the California Institute of Technology (See Chapter 4, Section 4.5.3).

○ While analysis methods were being developed, engineers needed additional knowledge about nonlinear behavior of structural components. Substantial testing of materials and connection assemblies to justify actual behavior were undertaken from 1950 through 1990 at numerous universities (University of California, Berkeley; University of Illinois; University of Michigan; University of Texas; etc.).

○ Since 1980 to the present, sophisticated computer analysis programs have been and continue to be developed to facilitate design of complex structural systems and the study of nonlinear behavior.

In the past 70 years substantial change and progress have taken place, not only in California but also over the entire United States, so that concepts and systems can now be utilized that previously could only be dreamt about.

## 7.3 HISTORIC AND CURRENT STRUCTURAL– SEISMIC SYSTEMS

### 7.3.1 Early Structural Systems–Pre-1906 San Francisco Earthquake

San Francisco in 1906 had a varied building stock with a few basic structural systems widely represented. Almost all common residential buildings were of light-frame wood construction, and most performed well in the earthquake, except for those on poor, weak soils or those with unbraced lower story walls. Most small- to medium-sized business buildings (about five to six stories in height) were constructed with brick masonry-bearing walls, using wood-framed floors and roofs. These buildings had a variable performance, with upper stories experiencing partial collapse and masonry walls typically showing shear cracks to varying degrees. Tall buildings, constructed during the previous 10 to 15 years (prior to 1906), utilized a steel frame to support gravity loads and provided unreinforced brick/stone perimeter walls which served to provide lateral load resistance. These buildings generally performed well during the earthquake. Most buildings, when subjected to the firestorm after the earthquake, did not do well.

The general conclusion following the 1906 earthquake was that a steel-framed building designed to support gravity loads and surrounded with well-proportioned and anchored brick walls to resist earthquake forces was a superior structural system, and it was commonly adopted by the design profession.

## 7.3.2 The Early Years (1906 – 1940)

Immediately after the 1906 earthquake, when reconstruction and new buildings became essential, a variety of new structural concepts were adopted. Brick masonry infill walls with some reinforcing were used, and steel frames were designed to carry lateral loads using one of the following ideas: knee bracing, belt trusses at floors to limit drift, rigid-frame moment connections using column wind-gussets, or top and bottom girder flange connections to columns.

As concrete construction became popular after 1910, concrete moment-frame buildings together with shear walls, emerged for industrial and lower height commercial buildings. Concrete slowly replaced brick as a structural cladding after 1930, and buildings commonly used a light steel frame for floor support with a complete perimeter concrete wall system for lateral loads.

## 7.3.3 The Middle Years (1945 – 1960)

Immediately after World War II, construction of large projects started again. New ideas were common, and some refinement of framing systems for tall buildings was proposed and adopted.

Expressive structural systems were studied and used, but they were usually covered from view with conventional exterior finishes.

The transition from riveted connections to high-strength bolted joints occurred in the 1950s. By 1960, another steel connection change was starting to occur; girder flanges welded directly to columns to create moment frame connections. Because engineers initially did not trust the limited use of moment frames, structural designs were conservative, with substantial redundancy created by utilizing complete moment-frame action on each framing grid, in each direction.

### 7.3.4 The Mature Years (1960 - 1985)

The 25-year period from 1960 to 1985 represents the "mature years", in that substantial projects were completed using the concepts of either ductile moment frames or concrete shear walls.

The structural engineering profession accepted the validity of 1) ductile concrete-moment frames, 2) ductile shear walls, or 3) ductile welded steel-moment frames as the primary structural system for resisting lateral loads. The primary design activity became optimization of the system or, in other words, how few structural elements would satisfy the minimum requirements of the building codes. Substantial connection tests were carried out at university laboratories to justify this design approach.

### 7.3.5 The Creative Years (1985 - 2000)

After the damage caused by the 1989 Loma Prieta earthquake (San Francisco Bay Area) and the 1994 Northridge earthquake (Los Angeles), the structural engineering profession began to ask itself about actual earthquake performance. Would real performance differ from the solution obtained by simple compliance with the building code? This investigative process defined many issues, and one of the most important was the dissipation of seismic energy by the building structure. The pursuit of this issue led engineers to the consideration of dual systems and seismic isolation to limit lateral displacement.

Several significant solutions have been developed using the dual-system concept with stable cyclic seismic behavior:

1. **Dual–system of steel moment frames and eccentric braced frames.** The more rigid eccentric brace provides primary stable cyclic behavior, while the moment frame provides good flexural behavior as a back-up system.

2. **The dual-system steel-moment frame and passive seismic dampers** provide high damping, which significantly reduces the seismic loads imparted to the moment frame.

3. **Unbonded steel braces** with the brace providing stable tension-compression behavior, a significant improvement over the conventional braced frame.

4. **Coupled 3-part systems with moment frames, links, and shear walls** to provide a progressive resistance system in which the resistance progresses from the most rigid system to the more ductile-flexible system.

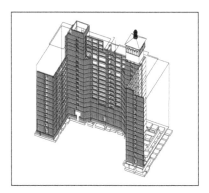

5. **Seismic isolation,** developed in the early 1980s, is a completely different and reliable concept, in which the building structure is supported on isolation bearings and is effectively separated from the ground, significantly reducing seismic response.

Each of these systems is part of an overall framing concept. These dual and stable mechanisms represent the current search for reliable seismic performance.

## 7.4 BACKGROUND AND PROGRESSION OF STRUCTURAL–SEISMIC CONCEPTS

The progression of seismic systems selected by structural engineers has resulted from three factors:

1.  Study of Past Earthquakes

    Learning from past earthquake performance: Successful seismic structural systems continue to be used; unsuccessful systems are eventually abandoned. New and better ideas frequently flow from observed earthquake damage.

2.  Research Data

    New ideas for structural concepts are frequently developed jointly by design engineers and university research laboratories. These systems are physically tested and analytically studied.

3.  Building Codes

    Finally, structural systems, that are listed in building codes eventually are used by many engineers as "approved". The problem with code concepts, in these times of rapidly changing systems, is that codes are created about 5 to 10 years after new ideas are developed, so that codes may no longer be current or at the cutting edge of new thinking; overly specific codes may tend to stifle and delay new ideas.

### 7.4.1 Development of Seismic Resisting Systems

Over a 100-year period, seismic resisting systems have developed substantially. The use of San Francisco buildings as a typical measure of the evolution provides a good summary of the past and present, and some indication of the future.

A summary of individual buildings gives a clue as to thinking. The plot of systems (Figure 7-1) connects the concepts and indicates the progression of ideas.

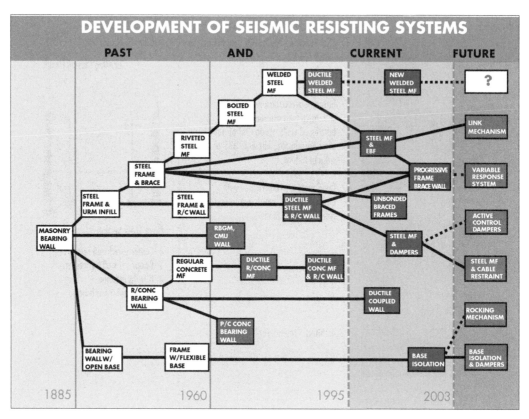

Figure 7-1: Development of Seismic Resisting Systems.

## 7.4.2 Pictorial History of Seismic Systems

The following pages provide a visual history of key features in the evolution of seismic systems developed and utilized in San Francisco and other Western regions of high seismicity.

### Old St. Mary's Church San Francisco
### 1869 - 1874

Strong tapered masonry tower, with nave built with unreinforced masonry buttressed walls. Wood roof of nave burnt after earthquake, but walls all stood without failure.

Good masonry construction with buttressed walls, which apparently resisted seismic loads.

photo: historic

Architect: Crane & England

existing steel roof trusses (1929)

existing masonry walls

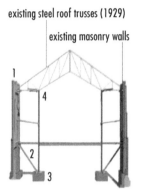

Nave Cross-Section

1 Center cored wall reinforcing
2 Basement wall buttress struts
3 New foundation
4 Roof diaphragm bracing

### Old U.S. Mint San Francisco
### 1869 - 1874

A massive masonry-bearing wall building with many thick interlocked walls. The walls benefited from continuous horizontal interlocked brick and a heavy horizontal diagonal rod bracing system at the attic level. The building performed well in the 1906 earthquake.

Architect: Alfred B. Mullet

Good masonry reinforcing provided continuity, and steel tension anchor rods provided an effective roof diaphragm.

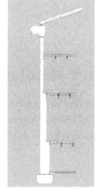

photo: unknown

### Light Timber Frame San Francisco 1880 - 1930

Wood framing provided good seismic performance, provided that the foundation system is stable and the lowest level is shear resistant and anchored to a strong foundation (not unreinforced brick). This was not the situation with this building.

Architect/Builder: Unkown

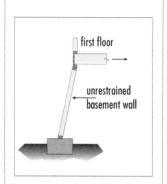

### U.S. Court of Appeals San Francisco 1902 - 1905/1931

A U-shaped, steel-braced frame building with small bays and substantial masonry infill. Concept worked well in 1906 earthquake, even though adjacent to a soft bay-mud soil type. An addition was added in the 1930s.

Architect: James Knox Taylor
Architect (Addition) George Kelham
Engineer: (Addition) H.J. Brunnier
Architect/Engineer (Upgrade 1996): Skidmore, Owings and Merrill

This structure was upgraded and restored utilizing a base-isolation solution, 1996.

base-isolation upgrade

### Mills Building San Francisco 1891 and 1909

Strong form, good steel framework, and good unreinforced masonry infill walls all combined for a seismic resistant building.

Architect: Burnham & Root, Willis Polk

Building was the tallest in San Francisco when it was constructed and demonstrated the seismic strength of steel working in conjunction with masonry infill.

photo: unknown

**Ferry Building  San Francisco
1895 - 1903**

Steel-framed building with masonry perimeter, large arched roof truss probably acted as an energy absorbing roof diaphragm spring. Tower was rod braced for wind loads.

The project has been recently seismically upgraded.

Architect: A. Page Brown / E. Pyle
Seismic Upgrade Architect: SMWM
Seismic Upgrade Structural Engineer: Rutherford/Chekene

Steel frame and details demonstrated its ability to absorb seismic energy.

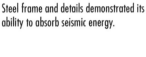

photo: Presidio Army Library

**St. Francis Hotel San Francisco
1904 - 1905**

Initial building was U-shaped steel frame with URM infill walls, which performed well during 1906 earthquake. A successful seismic design with structural steel frame and masonry infill.

Architect: Bliss & Faville

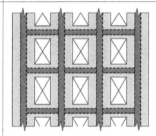

partial elevation

**Flood Building San Francisco
1904**

12-story wedge-shaped office building survived the 1906 earthquake and fire.

Architect: Albert Pissis

Another major steel frame that provided vertical support surrounded by brick masonry, which added energy absorption. This combination saved the monumental building in the 1906 earthquake.

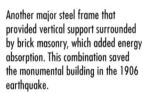

|  photo: Stanford University Archives | **Memorial Church Stanford University**  1906 earthquake damage shown. Additional damage in 1989. Seismic upgrade after 1906 and 1990.  Architect: Charles Coolidge & Clinton Day Engineer for 1990 upgrade: Degenkolb Engineers | Timber and masonry memorial church.  Large unrestrained walls collapsed. |
|---|---|---|
|  photo: Stanford University Archives | **Library Building Stanford University**  Serious damage, building demolished after 1906 earthquake.  Architect: Unknown | Steel-framed dome & drum with rigid masonry drum.  Masonry drum was too flexible and not restrained by masonry walls. |
|  photo: San Francisco Library | **Old City Hall San Francisco**  Serious damage; building demolished after 1906.  Architect: Augusta Laver | Steel-framed building with unreinforced masonry walls.  Steel-framed dome & braced drum with rigid masonry cladding.  Masonry drum was not restrained against "ovaling" during the earthquake and caused the brick cladding to separate from the steel frame. |
|  photo: San Francisco Library | **Downtown, Union Square San Francisco**  Buildings survived after 1906 and continue to function today  Architect: Various Architects | Union Square-steel framed buildings with masonry walls.  Bare steel buildings were under construction during the 1906 earthquake.  Good performance. |

### Royal Globe Insurance Building San Francisco 1907

A post-1906 earthquake building constructed with steel frame and reinforced brick perimeter infill. This was the first reinforced masonry facade in San Francisco after 1906.

Architect: Howells and Stokes
Engineer: Purdy and Henderson

Brick was reinforced apparently to prevent cracking and falling of URM.

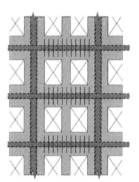

### City Hall San Francisco 1913 - 1915

A replacement for the original monumental City Hall which failed in 1906. The new City Hall was rapidly constructed to coincide with the 1915 Pan Pacific Exposition. Steel-framing connections were designed for gravity loads without moment connections to facilitate the rapid erection of steel. Virtually the entire seismic resistance was provided by infill perimeter masonry walls. The engineer also believed in a seismic softstory as a means of protecting the building, relying solely on the massive masonry perimeter walls with simple connections between embedded steel columns and girders.

Architect: Bakewell & Brown
Engineer: C.H. Snyder
Reconstruction Engineer:
Forell/Elsesser Engineers

City Hall suffered significantly in the 1989 earthquake and was retrofitted with a real "soft-story" using 530 elastomeric isolators at the base of structure.

SEISMIC DESIGN–PAST, PRESENT AND FUTURE

### Shreve Factory San Francisco 1908

An early reinforced concrete frame building, slender columns, minimum girders—a hazardous condition. Building was retrofitted in 1982 with new concrete shearwalls to protect the weak concrete frame.

Retrofit Architect: MBT Architecture
Engineer: Forell/Elsesser

---

### Typical Office Building San Francisco

A typical steel frame of the 1920s with belt-trusses used to limit lateral drift to the columns. The entire perimeter was encased in reinforced concrete for stiffness and fireproofing.

Architect: unknown

This system relied on the steel for seismic load resistance, and utilized the concrete walls for primary earthquake resistance.

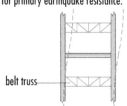

belt truss

A classic industrial-style concrete building designed to resist earthquakes by a combination of shear walls and slab-column frames, and to provide fire resistance.

---

### Hallidie Building San Francisco 1917

Concrete construction had been in general practice for about 10 years. This was an early reinforced concrete flat slab building, noted for its famous glass facade facing the street. This is a 3-sided perimeter shear-wall building with some moment-frame action between slab and columns.

Architect: Willis Polk

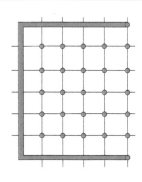

floor plan

---

**Standard Oil Building San Francisco
1922 / 1948**

A large steel frame designed with seismic knee braces, this U-shaped building was tall for the time, 25 stories. Cladding was infill URM, which provided additional drift

Architect: George Kelham
Engineer: H.J. Brunnier

A continuation of the 1906 framing concept with added steel bracing.

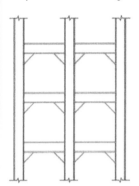

photo: H.J. Brunnier

photo: H. J. Brunnier

**Shell Oil Building San Francisco
1929**

A tall, slender 28-story steel frame structure, with steel wind gusset plate connections for lateral load resistance, together with concrete and unreinforced infill.

Architect: George Kelham
Engineer: H.J. Brunnier

A good example of wind-gusset plate joints for primary lateral resistance. The gussets provided enhanced moment-joint connections.

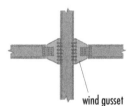

wind gusset

photo: H.J. Brunnier

### Russ Building San Francisco 1927

The tallest building in San Francisco at the time, with full riveted moment connections, encased in reinforced concrete and URM facing walls.

Architect: George Kelham
Engineer: H.J. Brunnier

A continuation of post-1906 construction but with steel acting together with reinforced concrete infill walls to resist seismic forces.

photo: H.J. Brunnier

### Mills Tower Addition San Francisco 1931

An addition to the successful 1891 original Mills Building. This 1931 addition utilized steel-moment frames created by top and bottom T-shaped flange plates with riveted connections. This type of moment joint was an early use of top and bottom tee-plates, which became standard for 30 years.

Architect: Willis Polk
Engineer: C.H. Snyder

An early use of full moment connections to provide the complete seismic capacity.

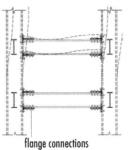

flange connections

photo: Bethlehem Steel

**Typical Apartment Building
San Francisco 1920 - 1930**

A slender steel framework designed
to support the concrete floor loads.
All lateral loads are resisted by the
reinforced concrete perimeter walls.

Architect: unknown

photo: Bethlehem Steel

photo: Bethlehem Steel

**Hoover Tower Stanford
1938**

A minimum steel frame designed as
initial supports for floors (similar to
apartment buildings), full lateral seismic
loads are resisted by a heavy reinforced
concrete perimeter wall.

Architect:  Bakewell & Brown

This building demonstrated the ultimate
use of the light steel frame with concrete
perimeter walls, acting to resist all wind
and earthquake loads.

**US Appraisers Building
San Francisco 1940**

A 12-story office building designed with
a heavy perimeter steel moment frame
utilizing belt-trusses at each floor, and
heavily reinforced concrete perimeter
walls for additional lateral capacity. An
unreduced base shear of 0.08W was used.

Architect/Engineer:
U.S. Government

photo: Degenkolb

### Park Merced Apartments
### San Francisco 1947

Cast-in-place L-shaped, 12-story concrete residential towers utilizing perimeter shear walls.

No steel frame.

Architect: L.S. Shultz
Engineer: John Gould

The beginning of full reinforced concrete-bearing and shear-wall buildings.

photo: Degenkolb

photo: San Francisco Library

### Standard Oil Building Addition
### San Francisco 1948

The second steel-frame addition required a substantial fabricated plate moment frame to accommodate the seismic forces created by the tall first story. The balance of the framework was a riveted moment frame utilizing top and bottom flange connections.

Architect: George Kelham
Engineer: H.J. Brunnier

photo: American Bridge

### Equitable Building
### San Francisco 1955

A unique steel moment-frame design utilizing the combined gravity and lateral load moment diaphragms to size and form the girders. Girder splices are at mid-span for shear only. The entire perimeter is encased in a substantial reinforced concrete wall. A lost opportunity to express the structure of a tall building, but a stout seismic solution.

Architect: Wilber Peugh
Engineer: Keilberg & Paquette

photo: American Bridge

photo: Bethlehem Steel

**Bethlehem Steel Building
San Francisco 1958**

A 14-story steel 2-way moment-frame structure without supplemental systems except for concrete fireproofing. All column-girder construction utilized a bottom flange girder haunch for drift control. Exterior columns offset from girders required a torsion box connection.

Architect: Welton Becket
Engineer: J.A. Blume

One of the last all-riveted structures.

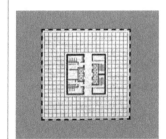

**Hancock Building
San Francisco 1959**

A 14-story reinforced concrete shear-wall structure with central core and perimeter pierced walls. An elegant reinforced concrete building with a symmetrical configuration.

Architect/Engineer:
Skidmore, Owings & Merrill

photo: Gould and Degenkolb

**American President Lines
San Francisco 1960**

A 22-story all-steel moment-frame building with large top and bottom T-section flange connections utilizing high-strength bolts. Joints were tested at the University of California, Berkeley.c

Architect: Anshen + Allen
Engineer: Gould & Degenkolb

A well-conceived steel frame with 2 directional frames on all grid lines.

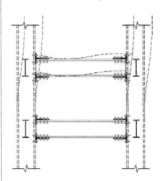

photo: H. J. Brunnier

**Zellerbach Building
San Francisco 1960**

A monumental 19-story steel-frame building with 60-foot clear span for the office wing, which is a 2-way moment-frame structure, and an adjacent connected braced service tower.

Architects: Hertzka & Knowles, SOM
Engineer: H.J. Brunnier

A steel moment-frame statement for the clear span offices, but with concrete walls around the braced service tower to add seismic strength and stiffness.

photo: H. J. Brunnier

Photo: Bethlehem Steel

### Alcoa Building San Francisco 1964

A building where a portion of the seismic bracing is expressed. The complete seismic system is dual: 1) the perimeter diagonal bracing, and 2) an interior moment-resisting 2-way frame.

Architect/Engineer:
Skidmore, Owings & Merrill

An elegant solution to express the structural seismic needs of the building.

Photo: Bethlehem Steel

Photo: H.J. Brunnier

### Bank of America Building San Francisco 1968

At 50 stories, the tallest, full-plate floor building in San Francisco, utilizing a 2-way grid of steel-moment frames with box column welded connections.

Architect: Skidmore, Owings & Merrill and Pietro Belluschi
Engineer: H.J. Brunnier

One of the last major 2-way moment-frame grid structures.

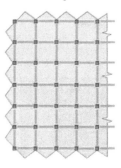

Photo: Bethlehem Steel

### Transamerica Tower San Francisco 1972

The tallest structure in San Francisco with its pyramid steel-framed form. A sloping moment frame above is supported on triangular pyramids towards the base. These pyramids transition to vertical frame columns at the lowest level.

Architect: William Pereira
Engineer: Chin and Hensolt

A unique and special structure designed as a "monument" for the Transamerica Corporation. The figure shows the base structure for columns.

Photo: Bethlehem Steel

SEISMIC DESIGN–PAST, PRESENT AND FUTURE

photo: H.J. Brunnier

**Golden Gateway Towers
San Francisco 1960**

25-story reinforced concrete dual shear-wall and moment-frame tower supported on massive drilled pier foundations. The first concrete tower blocks with low floor-to-floor elevations.

Architect: Skidmore, Owings & Merrill / Wurster Bernardi & Emmons / Demars & Reay
Engineer: H.J. Brunnier

The dual concrete shear walls and moment frames represented a new direction for San Francisco's high-rise buildings. This project was an early, ductile concrete moment frame.

**Summit Apartment Building
San Francisco 1968**

An expressive tower reflecting the seismic and gravity loads in the building form. Post-tension floor plates.

Architect: Neil Smith / Claude Oakland
Engineer: S. Medwadowski

A landmark concrete structure on top of Russian Hill.

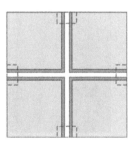

Floor Plan

**New St. Mary's Cathedral
San Francisco 1971**

A unique reinforced concrete shell form designed by Pier Luigi Nervi with an expressive structural form. A single structure-form solution.

Architect/Engineer (for shell):
Pier Luigi Nervi

A special hyperbolic shell solution that blends form with structure.

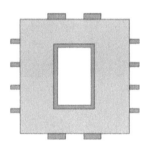

Roof Plan

**Pacific Medical Center
San Francisco 1970**

A tall, interstitial-space building designed as a moment frame and encased in a reinforced concrete shear-wall system.

Architect: SMP
Engineer: Pregnoff/Matheu

One of the "hospital systems" buildings developed in the 1970s.

photo: SMP

photo: BSD

**Loma Linda Hospital
Loma Linda 1975**

A hospital system building with a dual steel moment-frame/ concrete shear-wall seismic solution designed to resist real ground motions and minimize damage.

Architect: Building System Development and Stone, Maraccini and Patterson
Engineer: Rutherford & Chekene

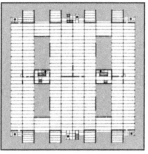

Plan

photo: Degenkolb

**UCSF Hospital
San Francisco 1978**

A substantial seismic demand for this site dictated steel-plated shear walls to provide adequate seismic strength.

Architect: Anshen + Allen
Engineer: Degenkolb

photo: Degenkolb

interior steel plate shear walls

boundary column splice

photo: Skilling, et al

### Bank of America Center
### Seattle 1965

50-story, 850,000 sf-in-area bank and office tower utilized a shear wall concept with 4 exterior walls designed with vierendeel trusses spanning to corner columns. At the 6th floor level, earthquake/wind loads are transferred to steel plate core walls.

Architect: NBBJ
Engineer: Skilling, et al.

A unique dual-perimeter and interior structural solution.

photo: Skilling, et al.

photo: Bethlehem Steel

### Metropolitan Life Building
### San Francisco
### 1973

A two-directional welded steel moment frame using shallow girder haunches at columns for drift control. Box columns for 2 way frame action.

Architect/Engineer:
Skidmore, Owings & Merrill

photo: Bethlehem Steel

### Marathon Plaza
### San Francisco 1985

A dual concrete seismic system with cast-in-place perimeter ductile moment frames and a cast-in-place shear-wall core. Precast panels clad the exterior, forming the exterior surface of the perimeter girder.

Architect: Whistler Patri
Engineer: Robinson Meier Juilly

A common, economical dual-system seismic solution, interior walls, exterior moment frame.

### Pacific Mutual San Francisco 1979

A perimeter welded steel-moment frame solution with light-weight GFRC cladding to minimize seismic mass.

Architect: William Pereira
Engineer: Chin & Hensolt

The use of deep but light-weight column sections to force yielding to occur in the girder sections.

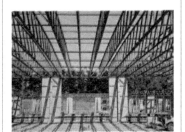

### Tomales High School Tomales 1980

A school "systems" building with custom-designed concrete cantilevered column-shear wall seismic elements.
Modular
light-gauge truss elements on a 5-foot

Architect: Marshall & Bowles
Engineer: Forell/Elsesser

Roof Plan

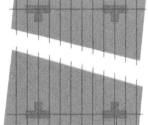

concrete bearing and shear walls

### Crocker Building San Francisco 1982

An early use of a perimeter-tube welded moment frame to achieve seismic resistance. No interior seismic resisting system.

Architect/Engineer:
Skidmore, Owings & Merrill

Photo: SOM

Photo: Skilling, et al (M-K)

**Bank of America Tower
Seattle 1982**

76-story, 2,000,000 sf in area, 945 ft. tall. Utilized visco-elastic dampers attached to exterior concret-filled super-columns to control drift and to reduce steel weight.

Architect: Chester L. Lindsey
Engineer: Magnusson - Klemencic

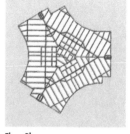

Floor Plan

Photo: Skilling, et al. (M-K)

**San Jose Federal Building
San Jose,1983**

An early use of steel eccentric braced frames with back-up welded moment frames to achieve economy and reliable seismic performance.

Architect: Hellmuth Obata & Kassabaum
Engineer: Forell/Elsesser

This building has experienced 3 moderate to strong earthquakes without damage.

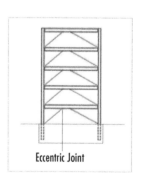

Eccentric Joint

Photo: N. Amin

### 333 Bush Street San Francisco 1985

An early version of a dual steel welded perimeter moment frame with a dual interior eccentric-braced frame.

Architect/Engineer:
Skidmore, Owings & Merrill

An excellent combination of systems, combining strength and drift control.

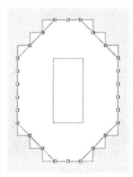

Plan

### Life Sciences Building UC Berkeley 1987

A unique reinforced concrete shear wall building with shear-links between shear walls in both longitudinal and transverse directions, designed to fracture during strong earthquakes. Two interior transverse walls and two exterior perimeter walls designed with diagonal bar cages at the links.

Architect: MBT Architecture
Engineer: Forell/Elsesser

This laboratory building was designed for vibration control and seismic damage control

| | | |
|---|---|---|
| <br><br>Photo: Bethleham Pacific | **San Francisco Armory**<br>**San Francisco 1920**<br><br>A 150-foot-span trussed arch with perimeter masonry walls. The arch system acts as a semi-rigid and flexible spring diaphragm.<br><br>Architect: William & John Woole | A conventional solution for long-span roof structures.<br><br>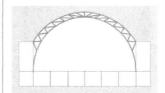 |
| <br><br>photo: unknown | **Cow Palace San Francisco 1938**<br><br>A unique and bold roof system with 3-hinged steel arches, supported on the ends of long steel truss cantilevers located at each side of the arena. The concrete perimeter side and end walls provide primary seismic resistance.<br><br>Architect: Engineer: Keilberg/Paqutte | The Cow Palace stretched the concept for the roof system by allowing pinned-jointed rotations of the 3-hinged arch to accommodate unplanned differential motions during seismic events.<br><br>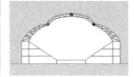 |
| <br><br>Photo: Robert Canfeld | **Haas Pavilion, UC Berkeley**<br>**Berkeley 1999**<br><br>A long-span truss expansion supported on seismic dampers.<br><br>Architect: Ellerbe Becket<br>Engineer: Forell/Elsesser | 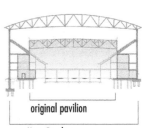<br>original pavilion<br>new Haas Pavilion |
| <br><br>Photo: SONM | **San Francisco International**<br>**Airport Terminal 1999**<br><br>Long-span double cantilever roof trusses with lower floors all supported on seismic isolators.<br>Architect: Skidmore, Owings & Merrill (Greg Hartman)<br>Engineer: Skidmore, Owings & Merrill (Navin Amin) | <br><br>Photo: SOM |

Photo: Robert Canfield

### Pacific Gas & Electric
### Historic Headquarters San Francisco
### 1925/1945 1995 Upgrade

The original steel-braced-frame PG&E Building, built in 1925 together with its neighbor, the Matson Building built in 1927, were damaged in the 1989 Loma Prieta earthquake. The strengthening solution joined the two together with a substantial 3-dimensional concrete shear wall and moment frame system designed to be a progressive resistance system, in which 3 elements (moment frames, shear links, and shear walls) yielded sequentially.

Architect: (original) Bakewell & Brown
Engineer: (original) C.H. Snyder
Rehabilitation Engineer: Forell/Elsesser

The new design solution provides basically the essential-facility performance required by PG&E.

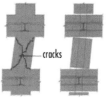

cracks

original          articulated facade

Photo: Robert Canfield

### UCSF Biomedical Research
### San Francisco 2003

This steel-braced frame utilized unbonded steel braces to achieve ductility with a conventional-braced seismic solution.

The unbonded brace provides equal tension and compression capacity.

Architect: Cesar Pelli & Associates
Engineer: Forell/Elsesser

Photo: Robert Canfield

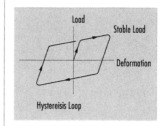

Load

Stable Load

Deformation

Hystereisis Loop

Photo: Robert Canfield

Photo: Robert Canfield

### State Office Building
### San Francisco 1999

After the 1994 Northridge earthquake, the State of California desired to minimize future damage costs. A dual special steel moment frame acting in conjunction with hydraulic dampers was developed. This system minimizes drift and protects the moment frames.

Architect: Skidmore, Owings & Merrill
Engineer: Forell/Elsesser Engineers

Damper Locations

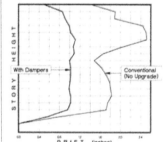

Floor Plan

### 911 Emergency
### Communications Center
### San Francisco 1998

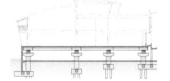

Photo: Robert Canfield

Seismic performance is critical for this 3-level, 911 Emergency Center. Seismic isolators significantly reduce seismic forces and interstory displacements, allowing a conventional steel braced frame to be used for the superstructure.

Architect: Heller & Manus/
Levy Design Partners Architecture
Engineers: Forell/Elsesser Engineers/)SDI

Photo: Robert Canfield

Seismic isolation for buildings is a relatively new concept. It effectively decouples the building from the shaking ground and significantly reduces the earthquake accelerations and interstory displacements.

## 7.5 COMMENTARY ON STRUCTURAL FRAMEWORKS

Both the steel-frame and concrete-frame buildings previously tabulated can be summarized on the basis of a primary framing system, considering both frameworks and structural cladding.

### 7.5.1 Steel Building Frameworks

It can be seen from the previous tabulation that steel frameworks have progressed from a simple steel frame, braced laterally by unreinforced masonry, to complete moment frames with full lateral load resistance. However, the 1994 Northridge earthquake in southern California created serious doubts as to the integrity of welded moment frames. Actually, several years before the 1994 earthquake, thoughtful structural engineers recognized the advantages of dual structural systems for the structural redundancy required to resist large earthquakes. The engineers had two separate tasks. The first task was to minimize the structure for economy, and the second task was to provide a secure load path to protect the structure.

A review of the basic eight steel-frame concepts tabulated in Figure 7-2 indicates that they are all reasonable, but some have substantially more redundancy than others:

1.  The simple **1890-1920** steel framework with unreinforced brick infill: The system has vertical support with an infill system that allows brick joint movement for energy dissipation. It is a good inexpensive system that allows for simple repair of brick after an earthquake.

2.  The **1910-1930** column-to-girder gusset-plate connection with nominally reinforced concrete infill walls: A good low-cost steel riveted detail with concrete providing stiffness for controlling lateral drift.

3.  The **1910-1940** trussed girder wind brace providing inexpensive drift control of the frame: The encasing concrete also provided substantial lateral stiffness, and forced the steel column sections to actually be stronger and stiffer and to create girder yielding, a good contemporary concept.

4.  The **1920-1940** knee-braced moment frame with concrete encasement provided a nominally stiff frame system.

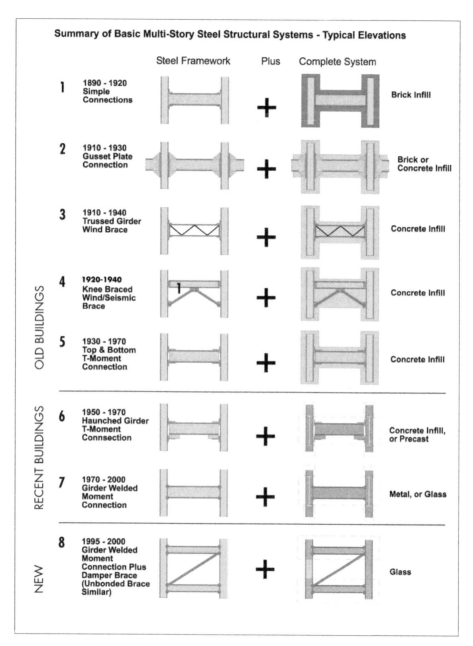

Figure 7-2:   Steel building frameworks.

5. The **1930-1970** riveted (or bolted) top and bottom girder fixity to the column created a steel-moment frame. The concrete fire-proofing encasement on buildings through 1960 enhanced the moment-frame stiffness.

6. The **1950-1970** top and bottom bolted haunched girder moment frame provided inexpensive girder stiffness and was especially strong if encased in cast-in-place concrete.

7. The more recent **1970-2000** all welded girder moment frame relied only on the steel system for seismic resistance and produced the most flexible of steel frames. These steel systems were not encased in concrete and were clad only with precast concrete, metal panels, or glass, not providing added structural strength or stiffness.

   After the Northridge earthquake, these conventionally welded frames were seen to be vulnerable, providing far less ductility than anticipated. A major FEMA-funded study has attempted to find solutions to this very significant problem. The current solutions tend to be expensive to fabricate.

8. The **1995-2000** steel-moment frames with a dual system of dampers, unbonded braces, or eccentric-braced frames, all clad with light-weight materials, appear to be good solutions.

The engineering profession had progressed fairly slowly until the early 1980s from the basic framing concepts that were first evolved in the early 1900s. When the concerns about seismic performance and energy dissipation became paramount, researchers and design engineers investigated mechanisms and configurations to supplement the basic rectangular grid framing in use for over 100 years.

### 7.5.2 Concrete Building Frameworks

Reinforced concrete framing became popular in California immediately after the 1906 earthquake. The significant framing systems that evolved are shown in Figure 7-3.

1. The early years, about **1908 to 1915**, featured nonductile frameworks of reinforced concrete column-girder moment frames. The earliest utilized wood floor infills between concrete floor girders, followed rapidly by all-concrete floors. It was not until the early 1960s that nonductile frames were recognized as an unsafe collapse mechanism, because of inadequate shear capacity, or because of poorly confined concrete.

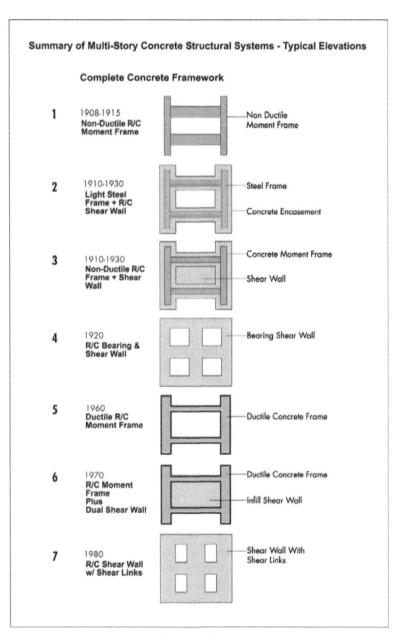

Figure 7-3:   Concrete building frameworks

2. Early steel frames for gravity load support were initially encased in masonry cladding, soon to be changed to reinforced concrete frame infill walls, **1910 - 1930**. This was a simple change which added definable strength around the steel.

3. Soon after the bare concrete frames were developed, they were infilled with concrete shear walls for lateral stability, **1910 - 1930**. Not all of these buildings were well conceived structurally, and because of the ease of adding or deleting walls, torsional problems became common for this type of building.

4. Significant reinforced concrete shear-wall buildings acting as bearing walls evolved in the**1920s**, without moment frames.

5. It was not until the late **1960s** and **early 1970s**, after substantial research at the University of Illinois, that the benefit of confined concrete columns and ductile concrete frames was recognized and adopted for dependable seismic resistance.

6. In the **1970s** and **1980s**, a dual system of ductile concrete moment frames coupled with confined concrete shear walls was recognized. This concept works best with a perimeter moment frame and an interior shear wall core, or the reverse—using a perimeter ductile wall and an interior ductile moment frame. Finally, a dependable concrete dual system evolved.

7. Shear walls coupled together with yielding shear links were developed in New Zealand in the **1980s** after successful tests carried out by Park and Paulay in 1975. This coupled-wall system is another excellent example of creative research used to develop a low-cost mechanism for seismic energy dissipation.

## 7.6 SYSTEM CHARACTERISTICS

Selecting a good structure requires engineering common sense. Common sense requires understanding the earthquake motion and its demands, and understanding the structural behavior of the individual systems available. There are differences of scale (small versus large), differences between elastic and inelastic behavior; and differences of dynamic responses and seismic energy dissipation. Structural and architectural configurations (such as regular versus irregular forms) are also significant in the performance. The many variables often make it difficult to select an appropriate system. The building code lists numerous structural systems, but it does not provide guidance in the selection of a system, and the many systems are not equal in performance.

Key performance issues are elastic behavior, inelastic behavior, and the related cyclic behavior resulting from pushing a structure back and forth. This behavior should be stable, nondegrading, predictable, and capable of dissipating a large amount of seismic energy.

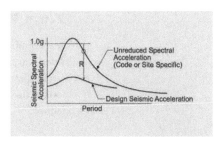

Figure 7-4: Design verses real acceleration.

### 7.6.1 Elastic Design—Linear Systems

The simple building code approach to seismic design requires diminishing an acceleration spectra plot by use of an R value. Elastic design is expressed by an R value, which is used to modify the acceleration spectral value to a simple seismic design force (Figure 7-4). This is a simple but frequently questionable method. It does not consider performance, nonlinear cyclic behavior, or most important—energy dissipation.

### 7.6.2 Post-Elastic Design—Nonlinear Drift

Inelastic design is a better indication of realistic lateral drift or deflection that results from real earthquake motions (Figure 7-5). Nonlinear drift impacts structural and nonstructural behavior. For significant seismic energy dissipation, the drift should be large, but for favorable nonstructural or content behavior this drift should be small. A building with a large but unstable structural drift will collapse. A building with a limited or small structural drift generally will not dissipate significant seismic energy without significant damage.

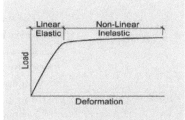

Figure 7-5: Elastic and inelastic.

Figure 7-6: Cyclic behavior.

### 7.6.3 Cyclic Behavior

A good measure of seismic performance is stable cyclic hysteretic behavior. The plot of load vs. deformation of an element, for motion in both directions, represents cyclic behavior (Figure 7-6). If the load curves are full, undiminished, without "necking down", they represent a stable system that is ductile and has sufficient capacity to deliver a constant level of energy dissipation during the shaking imposed by an earthquake. Degrading cyclic systems may be acceptable if they degrade slowly and in a predictable manner.

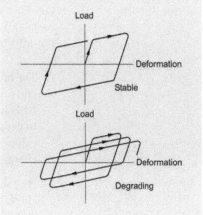

## 7.6.4 Performance-Based Seismic Design

Years of physical testing and corresponding analytical studies have provided a greatly increased understanding of earthquakes, materials and assemblies. Recent seismic performance design has forced engineers to look at the full range of structural behavior, from linear to nonlinear to failure.

Good, dependable seismic systems behave with a stable, nondegrading behavior; others do not. The ones that are not stable diminish in capacity with continuing cycles, until they have little capacity to dissipate seismic energy. If the earthquake continues for a long duration (over 10 to 15 seconds), the structure can become weak and unstable, and may fail.

The single conventional system with only one means of resisting seismic forces is especially vulnerable to long-duration earthquakes, because when that system degrades, no alternative exists. This concern has evolved for high-performance structures into the concept of utilizing multiple-resisting structural systems that act progressively, so that the overall structural capacity is not significantly diminished during the earthquake.

## 7.6.5 Nonlinear Performance Comparisons

Considering only strength and displacement (but not energy dissipation), six alternative structural systems are represented by seismic performance "pushover" plots for comparison for a specific four-story building, but a useful example of a common low- to mid-rise building type (Figure 7-7). For other building heights and types, the values will vary. Each system was designed and sized for code compliance, all with a lateral drift of less than one inch. However, the nonlinear displacement (or drift) for each alternative system is significantly different:

○ The **braced frame** (BF) needs to reach the 5% damped spectral curve, but the system fails at a drift of about one-and-a-half inches, far below the target of 5%t damping. A possible solution would be to design the brace for about four times the code value; excessively expensive, this solution deals only with strength, not with seismic energy.

○ The **shear wall** (SW) needs to reach the 10% damped or spectral curve, but can only reach a drift of about two inches, about half of the required drift. Again, significant over-design might solve the strength problem, but may not solve the energy issue.

○ The **eccentric braced frame** (EBF) needs to reach the 10% damped curve at a drift of about 5.5 inches, which exceeds its ductility and would create excessive deformation. The solution is to add more capacity.

○ The **steel-moment frame** (MF) without a supplemental system needs to reach a 5%t damped curve at a drift of about nine to ten inches, far in excess of its capacity. The only solutions are to over-design the MF or to add a supplemental system.

○ The **steel-moment frame with passive dampers** (MF + Dampers) needs to reach the 20% damped spectra curve which it does at a drift of about four inches. This is a reasonable solution.

○ The **base-isolation** solution (BI) satisfies the demand at a drift from 10 to 12 inches, which it can easily do without damag,e because most of the lateral drift occurs in the isolation itself.

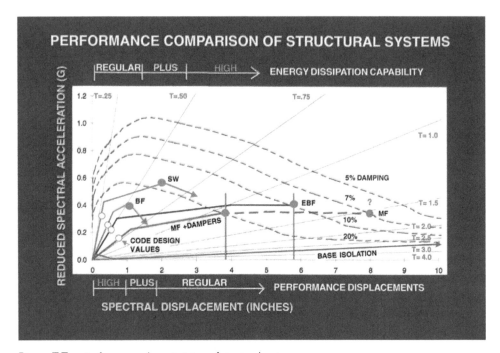

Figure 7-7:   Performance characteristics  of structural systems.

### 7.6.6 Energy Dissipation

During the duration of ground shaking, seismic energy is "stored" in a structure. If that input energy can be stored and dissipated without damage to the structure, the behavior is favorable. If, however, the structure cannot store the energy, the system may rupture with either local or catastrophic behavior (Figure 7-8A). Earthquake energy accumulates during ground shaking so that the total energy continually increases. If the structure can safely dissipate energy as rapidly as it is input to the structure, no problem occurs. In fact, the structure can store some of the energy due to ground shaking, and safely dissipate the balance after the earthquake stops. If the structure cannot keep up with the energy input, the structure may suffer minor cracking, or rupture, or major failure (Figure 7-8B).

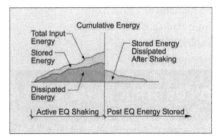

Figure 7-8A:   Energy stored.                    Figure 7-8B:   Energy storage failure.

### 7.6.7 The Design Approach—Force vs. Energy

Searching for the perfect system with conventional solutions has been limited to date, because seismic forces are used that are based on accelerations, and then the resulting lateral drift or movements of the structure are reviewed to check the behavior of both structural and nonstructural elements, such as rigid cladding. This process has an inherent conflict.

Seismic accelerations are used because a design force, **F**, from the relation **F=Ma**, can easily be obtained where **M** is the building mass, and **a** is the ground acceleration.

For the typical stiff-soil site, the larger the acceleration **a**, the larger the seismic force **F**. The larger the force **F**, the stronger the structure is; the stronger the structure, the stiffer the structure. The stiffer the structure is, the higher the seismic acceleration, and so on. A strong, stiff structure appears to be good conceptually. The lateral drift (or horizontal displacement) is at a minimum, which is also good. However, the contradiction of this approach comes from seismic energy dissipation, which is the fundamental need and characteristic of good seismic design; energy dissipation comes from large displacements, not small displacements.

**Large displacements** are needed to dissipate seismic energy, and **small displacements** are needed for lateral drift control to protect cladding, glazing, and interior systems. This produces a conflict for most of the classic conventional structural concepts and normal nonstructural components.

The most useful **seismic systems** are those that have predictable, stable, nondegrading cyclic behavior. Contemporary structures with these characteristics are base-isolation systems; moment frames with dampers; shear-link systems, such as coupled shear walls and eccentric-braced frames; and other dual-resistance systems with built-in redundancy.

The most useful **nonstructural components** are those that accommodate large lateral movements without failure.

## 7.7 THE SEARCH FOR THE PERFECT SEISMIC SYSTEM

The 100-year review of seismic systems indicates a slow development of structural solutions. In fact, most new development occurs after a damaging earthquake. Perhaps this slow periodic development is due to the need to discern whether previous ideas were successes or failures. The following topics represent steps and ideas in the search for the "perfect system."

### 7.7.1 Structural Mechanisms

There have been periodic significant conceptual breakthroughs in structural thinking. For example, Dr. Ian Skinner and his team, working at the New Zealand National Laboratory, developed in 1976 a set of energy dissipating concepts suitable for use in seismic protection. Among the most notable concepts was the practical use of elastomeric isolation bearings for global protection of complete structures. Some of the other concepts

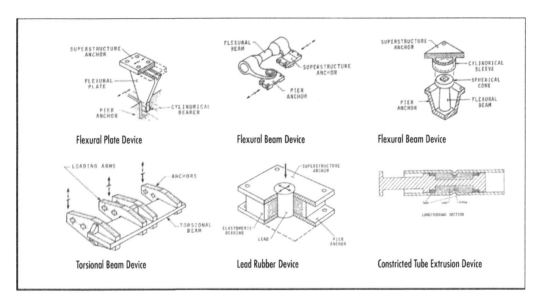

Figure 7-9: A set of energy dissipating devices.

SOURCE: NEW ZEALAND NATIONAL LABORATORIES

developed in New Zealand have also been utilized in building design. Figure 7-9 illustrates the set of mechanisms published in 1980.

Many of these relatively simple mechanisms were new to the design professions and to constructors; consequently, they represented unknown processes and unknown costs. Development and education are the keys to the acceptance and adoption of these systems. Perhaps a few well-publicized, prototype projects would familiarize the design professions and constructors with the details and costs of these good ideas. To date, this has only been done with the flexural plate and the lead-rubber isolation bearing.

## 7.7.2 Semi-Active and Active Dampers

Recent research and design work in Japan and the United States has focused on approaches for structural control during actual wind storms and earthquakes. These approaches have been summarized by Hanson and Soong, and can be divided into three groups: **passive systems**, such as base isolation and supplementary energy dissipation devices; **active**

**systems** that require the active participation of mechanical devices whose characteristics are used to change the building response; and **hybrid systems** that combine passive and active systems in a manner such that the safety of the building is not compromised, even if the active system fails. The current systems being studied are characterized by devices that control properties and behaviors. Some of these systems are in limited use; others are still in development.

The goal of these devices is to respond actively to the variable character of wind and earthquake displacements, velocity, accelerations, etc., by adding damping or altering stiffness. This controlled behavior will provide the needed resistance to respond to ever-changing earthquakes. However, the challenge with semi-active or active control systems is maintaining their behavior trouble-free over an extended period of time - specifically over many years. These concepts, although very appealing, will require some time to perfect and bring to market.

## 7.7.3 Cost-Effective Systems

The most important measure of good earthquake-resistant design is the impact on the structure after the earth has stopped shaking. With little building damage, repair costs are low. With significant damage or collapse, repair or replacement costs are high. The measure of success in seismic design is selection of a structure that will suffer minimum damage, and with corresponding low post-earthquake repair cost. The behavior of each structural system differs with earthquake ground motions, soil types, duration of strong shaking, etc. Our past observations of damage yields the best measure of future repair costs. Systems with stable cyclic behavior, good energy dissipation, and controlled interstory drift will yield low repair costs (Figure 7-10).

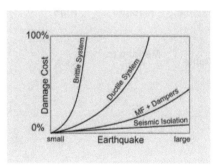

Figure 7-10

System performance.

**Seismic - Structural System Characteristics**

| System | | Seismic Design R | Non-Linear Drift | Cyclic Behavior | Energy Dissapation | Post EQ Repair Cost |
|--------|--|------------------|------------------|-----------------|--------------------|--------------------|
| UR Masonry Walls | 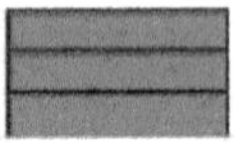 | 1.5 | Medium to Collapse | Unstable | Low | High |
| Timber Framing | 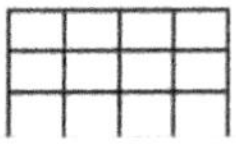 | 5.5 | Medium | Stable | Medium to High | Medium |
| Steel Frame URM Wall | 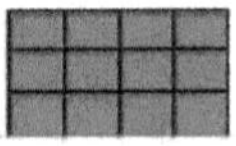 | (3.5) | Medium | Stable | Medium | Medium |
| Steel Frame + R/C Walls | 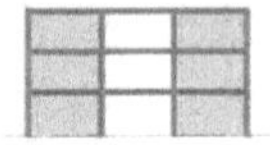 | 5.5 | Medium | Stable | Medium | Medium |
| Non-Ductile R/C - MF | 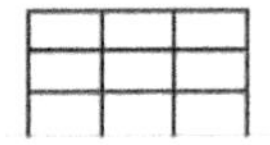 | 3.5 | Large to Collapse | Unstable | Low | High |
| Steel Frame + Braces | 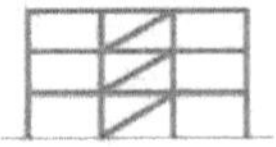 | 4.5 | Large to Collapse | Unstable | Low | High |
| R/C Shear Walls |  | 4.5 | Medium | Stable to Unstable | Medium to High | Medium to High |
| RG Masonry Walls | 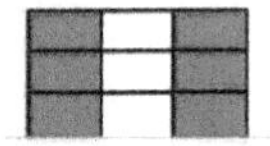 | 4.5 | Medium to Large | Stable to Unstable | Medium | Medium to High |
| Non-Ductile Steel MF | 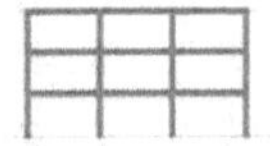 | 4.5 | Medium to Large | Unstable | Medium | High |
| Composite Timber & Steel | 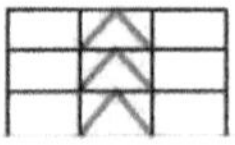 | 4.4 | Medium | Stable | Medium to High | Low |

Figure 7-11A: Structural seismic characteristics.

## Seismic - Structural System Characteristics

| System | | Seismic Design R | Non-Linear Drift | Cyclic Behavior | Energy Dissapation | Post EQ Repair Cost |
|---|---|---|---|---|---|---|
| RG Masonry R/C MF | 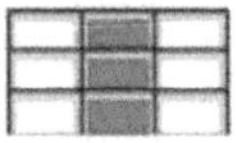 | 5.5 | Medium | Stable to Semi-Stable | Medium to High | Low to Medium |
| Ductile Steel MF | 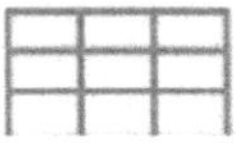 | 8.5 | Medium to Large | Stable to Semi-Stable | Medium | Medium to High |
| Ductile R/C MF | 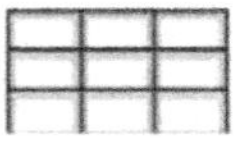 | 8.5 | Medium to large | Stable to Semi-Stable | Medium | Medium to High |
| Steel Eccentric Brace | 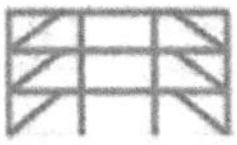 | 7.0 | Low to Medium | Stable | Medium to High | Low to Medium |
| R/C Link Shearwall | 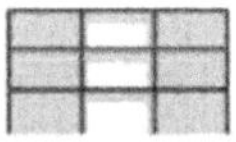 | 6 | Medium | Stable | Medium to High | Low to Medium |
| Dampers + Steel MF | 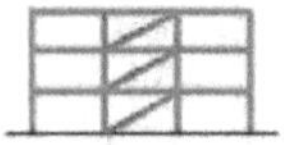 | 8.5 | Low to Medium | Stable | Very High to High | low |
| Unbonded Steel Brace | 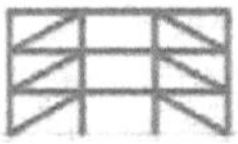 | 6.5 | Medium | Stable | Medium to High | Low to Medium |
| Rocking System | 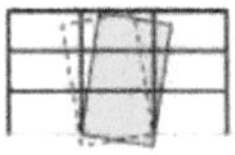 | (8.5) | Large Rocking Motion | Stable | High | Low |
| 3-Part Progressive System | 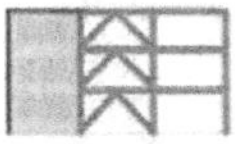 | 8.5 | Low to Medium | Stable | High | Low |
| Seismic Isolation | 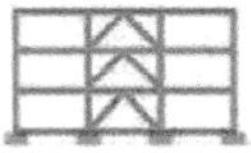 | (1.6-2.0) | Low Interstory Drift | Stable | Very High | Very Low |

Figure 7-11B: Structural seismic characteristics.

The above five attributes, which are tabulated in Figures 7-11A and 7-11B for the various significant seismic structural systems, show that a range of possibilities is indicated. These tables are presented for discussion, and the evaluations stated are the author's opinion, based on observations, analysis, and common goals for seismic performance. The green colored boxes are favorable conditions; the red tend to be unfavorable. The favorable structural systems will do the following:

○ Possess stable cyclic behavior

○ Control lateral drift

○ Dissipate seismic energy without failure

○ Create a low post-earthquake repair cost

The design reduction value R, discussed above in section 7.6.1, does not necessarily correlate with performance. The "R" value was a consensus value developed for conventional elastic design. With the advent of performance design based on nonlinear evaluation the R value serves only as a "rough estimate" of system behavior, but not a realistic estimate of performance.

### 7.7.4 Avoiding the Same Mistakes

Architects and engineers learn from their detailed investigations of past earthquake damage and can document the significant issues and lessons that can be learned about particular structural problems. Some problems occur because of inappropriate building or structural configuration, some because of brittle, nonductile structural systems, some because the building or structure could not dissipate sufficient seismic energy, and some because of excessive loads caused by dynamic resonance between the ground shaking and the building. Figure 7-12 illustrates some classic problems.

Why, with all our accumulated knowledge, does all this failure continue? Buildings tend to be constructed essentially in the same manner, even after an earthquake. It takes a significant effort to change habits, styles, techniques and construction. Sometimes bad seismic ideas get passed on without too much investigation and modification.

Figure 7-12: Examples of typical earthquake damage.

The challenge for architects, engineers, and constructors is to ask why so much damage occurred, and what can be done to correct the problem. Understanding the basic seismic energy demand is a critical first step in the search for significant improvement. Not all structural systems, even though they are conventional and commonplace, will provide safe and economical solutions.

### 7.7.5 Configurations Are Critical

Configuration, or the three-dimensional form of a building, frequently is the governing factor in the ultimate seismic behavior of a particular structure. Chapter 5 covers conventional configuration issues where conventional rectangular grids are used in building layout, design and construction. However, contemporary architectural design is changing, in large part because the computer allows complex graphic forms and analyses to be generated and easily integrated into a building design.

The resulting irregular, random, free-form grids and systems have just begun to be explored from the structural engineering viewpoint. They are frequently rejected because of various cost issues, and because of un-proven real earthquake behavior.

The potential for optimizing seismic resistance with respect to structural configuration is an obvious direction for the future. Structural form should follow the needs. How can we define seismic needs?

Buildings must dissipate energy; how does one configure a structure to dissipate energy? There are natural forms and design concepts that act as springs, rocking mechanisms, flexural stories, yielding links, articulated cable-restrained configurations, pyramid forms, cable anchors, etc. Any system that can dissipate seismic energy without damage is a candidate. Figure 7-13 illustrates some special concepts utilizing building configuration to dissipate energy.

### 7.7.6 Common-Sense Structural Design-Lessons Learned

The simple way to reduce seismic demand within a structure is to understand the actual demand. An earthquake is a dynamic phenomenon with all its classic characteristics. If one can reduce the effective damaging character of the earthquake, the behavior of a structure or building will be significantly improved. The following five issues can significantly reduce earthquake damage and related costs.

Figure 7-13: Concepts that use building configurations to dissipate energy.

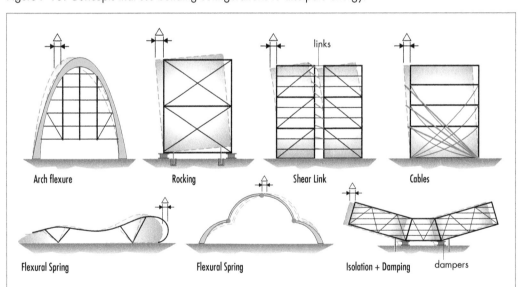

● Select the Appropriate Scale

The size and scale of the building should determine the appropriate structural solution. It is common sense to use a light braced frame for a small structure, such as one- or two-story wood or light-gauge steel systems. The light seismic system is compatible with the light building mass. As the building mass increases with the use of heavy concrete floors, the mass increases by a factor of 8 to 12 times. This would imply a much heavier seismic frame or brace (Figure 7-14), 8 to 12 times stronger per floor, and if the building grows from two to ten stories, the total mass increases from two to 80-120 times.

The same frame or brace which works well at two stories no longer works well at ten stories, because structural components when simply scaled up no longer behave in the same fashion. For instance, a light steel section with 3/8-inch flanges is ductile, but no longer has the same ductility when the flanges are two inches thick. The thick sections and welds are now subjected to high shears and are prone to failure.

Some structural systems are very forgiving at small sizes but not at a large scale. Alternative structural systems must be used for the larger, more demanding buildings. The appropriate systems must not degrade in strength and should have ample sustainable damping. It is critical that scale be considered, even though it is not considered by the seismic building codes. This scale issue requires study, observation, and common sense, but the issue of appropriate scale continually confronts designers. Careful, unbiased research is necessary.

● Reduce Dynamic Resonance

It is important to significantly reduce the dynamic resonance between the shaking ground and the shaking building, and to design the structure to have a period of vibration that is different from that of the ground period. The difference is simply illustrated with a classic resonance curve (Figure 7-15).

The relationship between the building period and spectral acceleration of a real earthquake varies between a site with firm soil and a site with soft soil (Figure 7-16).

If the building has a period of vibration, T1, corresponding to the peak ground acceleration, the most severe demand occurs. Shift the period,

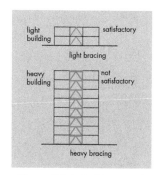

Figure 7-14: Building scale.

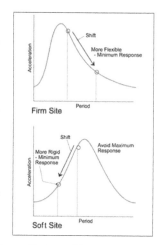

Figure 7-15: Building resonance.

Figure 7-16: Effect of resonance.

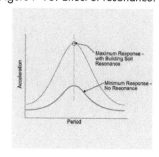

T2, by altering the structural system to lengthen the period (firm soil) or shorten or lengthen the period T2 (soft soil).

This requires design effort by both the architect and engineer, but the result can be a significant reduction in the seismic response.

### ● Increase Damping

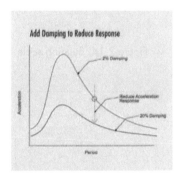

Figure 7-17: Reduction of demand by damping.

It is valuable to significantly increase the structural system damping. Damping reduces vibration amplitude similar to the hydraulic shock absorbers in an automobile, and damping reduces the structural demand (Figure 7-17), as illustrated in the spectral acceleration plot.

Significant damping can be introduced into a structure by 1) adding a passive damping system (hydraulic, friction, etc.); 2) utilizing a fractured concrete member such as a shear-link serving to couple two walls; 3) fracturing a concrete shear wall; 4) utilizing a seismic isolation system (with 10 to 20% damping), or 5) utilizing a tuned-mass damper (a challenging solution for most buildings). An increase of damping by using a non-distinctive system is a most positive solution for reducing seismic demand.

### ● Provide Redundancy

It is also important to add redundancy or multiple load paths to the structure to improve seismic resistance. After experience in many earthquakes and much study and discussion, the engineering profession has generally concluded that more than a single system is the ideal solution for successful seismic resistance. If carefully selected, multiple systems can each serve a purpose; one to add damping and to limit deflection or drift, the other to provide strength. Multiple systems also serve to protect the entire structure by allowing failure of some elements without endangering the total building.

An informative sketch of a classic redundant-framing concept, with frames on each grid line, versus a contemporary multiple system with two types of framing, one for strength, the other for damping, is shown in Figure 7-18. The current dual systems now being developed and utilized are a significant improvement over the historic single seismic resisting systems.

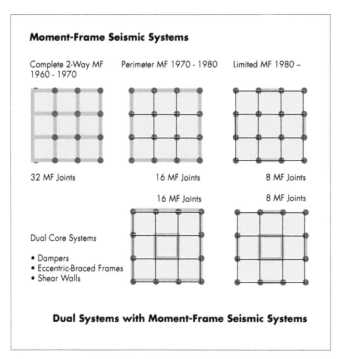

Figure 7-18: Single and multiple system concepts.

● Energy Dissipation

Solving the seismic energy dissipation problem is the ultimate test of good earthquake-resistant design. Since large building displacement is required for good energy dissipation, while minimum displacement is required to protect the many brittle non-structural components in the building, only one seismic resisting system adequately solves both aspects of this problem—seismic base isolation.

However, seismic isolation is not suited for all building conditions: specifically, tight urban sites adjacent to property lines (a movement problem), and tall buildings over 12 stories (a dynamic resonance problem). Cost of construction is another consideration (about a 1 to 2% premium). Seismic isolation is an ideal solution for irregular buildings and unusual or creative building forms that are difficult to solve with conventional structural systems.

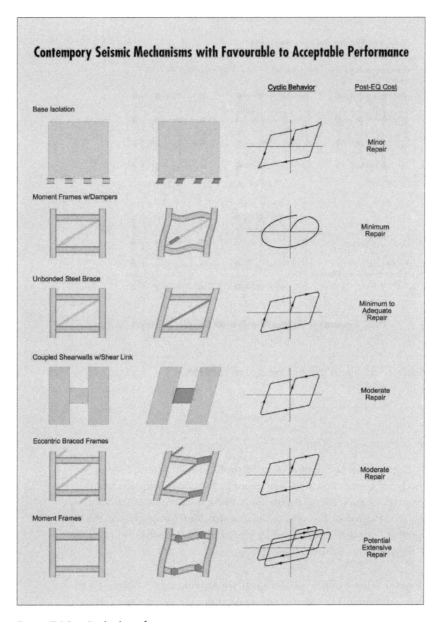

Figure 7-19: Six high-performance seismic structures.

Other dual systems can also solve the energy dissipation problem. Six seismic systems are outlined in Figure 7-19. Each system has a cyclic load-deformation loop which is full and nondegrading. Most importantly, each has a predicted minor-to-moderate post-earthquake damage, with a correspondingly low post-earthquake repair cost. This fact makes these good seismically sustainable buildings, expanding the meaning of "green building".

## 7.8 CONCLUSIONS

We must continue to develop and use increasingly realistic analysis methods to design new buildings and to modify existing structures. This is not a simple task because of the numerous variables: duration of shaking, frequency content of the seismic motion, displacement, velocity, acceleration, direction of the earthquake pulse, proximity to the active fault, and soil amplification effects. In addition to the ground effects, the structure has its own variables: size; shape; mass; period of vibration; irregularities in shape, stiffness, and strength; and variation in damping.

After the past 100 years of seismic design, the perfect architectural/structural solution is still elusive, but with creative thought, testing, and computers, the problem is more transparent, and we can now more rapidly study variables than in the recent past. Design work today has become part research, part invention, and common sense, and we are on the threshold of creative breakthrough.

## 7.9 REFERENCES

1. Park, R. and Paulay, T, 1975. *Reinforced Concrete Structures*, Wiley—Interscience.

2. Paulay, T. and Priestly, N, 1992. *Seismic Design of Reinforced Concrete and Masonry Buildings*, Wiley.

3. Hanson, R. and Soong, T.T, 2001. *Seismic Design with Supplementary Energy Dissipation Devices*, Earthquake Engineering Research Insitute, Oakland CA.

# EXISTING BUILDINGS–EVALUATION AND RETROFIT  8

by William Holmes

## 8.1 INTRODUCTION

It is widely recognized that the most significant seismic risk in this country resides in our existing older building stock. Much of the country has enforced seismic design for new buildings only recently; even on the West Coast, seismic codes enforced in the 1960s and even into the 1970s are now considered suspect. Although there are sometimes difficulties in coordinating seismic design requirements with other demands in new construction, the economical, social, and technical issues related to evaluation and retrofit of existing buildings are far more complex.

### 8.1.1 Contents of Chapter

This chapter describes the many issues associated with the risk from existing buildings, including common building code provisions covering older buildings, evaluation of the risks from any one given building and what levels of risk are deemed acceptable, and methods of mitigation of these risks through retrofit. A FEMA program to provide methods to mitigate the risk from existing buildings has been significant in advancing the state of the art, and this program is described in some detail, particularly the model building types used in most, if not all, of the FEMA documents.

### 8.1.2 Reference to Other Relevant Chapters

The basic concepts used for seismic design or estimation of seismic performance are the same for any building. Thus, the principles described in Chapters 4, 5, and 7 are applicable for older, potentially hazardous buildings. The development of seismic systems as seen through examples of buildings in the San Francisco Bay Area is particularly relevant to the issues covered in this chapter, because systems typically evolved due to poor performance of predecessors.

Nonstructural systems in buildings create the majority of dollar loss from buildings in earthquakes, although the quality of structural performance affects the level of that damage. Seismic protection of nonstructural systems, both for design of new buildings and for consideration in older buildings, is covered in Chapter 9.

## 8.2 BACKGROUND

In every older building, a host of "deficiencies" is identified as the state of the art of building design and building codes advances. Code requirements change because the risk or the expected performance resulting from the existing provisions is deemed unacceptable. Deficiencies are commonly identified due to increased understanding of fire and life safety, disabled access, hazardous materials, and design for natural hazards. Thus, it is not surprising that many of the older buildings in this country are seismically deficient, and many present the risk of life-threatening damage. It is not economically feasible to seismically retrofit every building built to codes with no or inadequate seismic provisions, nor is it culturally acceptable to replace them all. These realities create a significant dilemma: How are the buildings that present a significant risk to life safety identified? How is the expected performance predicted for older buildings of high importance to businesses or for those needed in emergency response? How can we efficiently retrofit those buildings identified as high risk?

The term **seismic deficiency** is used in this chapter as a building characteristic that will lead to unacceptable seismic damage. Almost all buildings, even those designed to the latest seismic codes, will suffer earthquake damage given strong enough shaking; however, damage normally considered acceptable should not be expected in small events with frequent occurrence in a given region, and should not be life threatening. Damage may be judged unacceptable due to resulting high economic cost to the owner or due to resulting casualties. Therefore, conditions that create seismic deficiencies can vary from owner to owner, from building to building, and for different zones of seismicity. For example, unbraced, unreinforced brick masonry residential chimneys are extremely vulnerable to earthquake shaking and should be considered a deficiency anywhere that shaking is postulated. On the other hand, unreinforced brick masonry walls, infilled between steel frame structural members, are expected to be damaged only in moderate to strong shaking and may not be considered a deficiency in lower seismic zones. Seismic deficiencies identified in this chapter generally will cause premature or unexpected damage, often leading to threats to life safety, in moderate to strong shaking. Buildings in regions of lower seismicity that expect Modified Mercalli Intensity (MMI) levels of not more than VII, or peak ground accelerations (PGA) of less than 0.10g (g = acceleration of gravity), may need special consideration.

Of course, every building with completed construction is "existing." However, the term **existing building** has been taken to mean those buildings in the inventory that are not of current seismic design. These groups of buildings, some of which may not be very old, include buildings with a range of probable performance from collapse to minimal damage. In this chapter, the term "existing building" is used in this context.

## 8.2.1 Changes in Building Practice and Seismic Design Requirements Resulting in Buildings that are Currently Considered Seismically Inadequate

Chapter 7 documents in detail how building systems have evolved in the San Francisco Bay Area. This evolution was probably driven more by fire, economic, and construction issues than by a concern for seismic performance, at least in the first several decades of the twentieth century, but many changes took place. Similarly, Chapter 6 gives a brief history of the development of seismic codes in the United States. It is clear that for many reasons, building construction and structural systems change over time. In the time frame of the twentieth century, due to the rapid increase in understanding of the seismic response of buildings and parallel changes in code requirements, it should be expected that many older buildings will now be considered seismically deficient.

Seismic codes in this country did not develop at all until the 1920s, and at that time they were used voluntarily. A mandatory code was not enforced in California until 1933. Unreinforced masonry (URM) buildings, for example, a popular building type early in the twentieth century and now recognized as perhaps the worst seismic performer as a class, were not outlawed in the zones of high seismicity until the 1933 code, and continued to be built in much of the country with no significant seismic design provisions until quite recently. Figure 8-1 shows an example of typical URM damage. The first modern seismic codes were not consistently applied until the 1950s and 1960s, and then only in the known regions of high seismicity. Of course, not all buildings built before seismic codes are hazardous, but most are expected to suffer far more damage than currently built buildings

Even buildings designed to "modern" seismic codes may be susceptible to high damage levels and even collapse. Our understanding of seismic response has grown immensely since the early codes, and many building

Figure 8-1: Example of buildings with no code design. Damage shows classic URM deficiencies.

characteristics that lead to poor performance were allowed over the years. For example, concrete buildings of all types were economical and popular on the West Coast in the 1950s and 1960s. Unfortunately, seismic provisions for these buildings were inadequate at the time, and many of these buildings require retrofit. Highlights of inadequacies in past building codes that have, in many cases, created poor buildings are given below.

● Changes In Expected Shaking Intensity and Changes in Zoning

Similar to advancements in structural analysis and the understanding of building performance, enormous advancements have been made in the understanding of ground motion, particularly since the 1950s and 1960s. The seismicity (that is, the probability of the occurrence of various-sized earthquakes from each source) of the country, the likely shaking intensity from those events depending on the distance from the source and the local soil conditions, and the exact dynamic nature of the shaking (the pattern of accelerations, velocities, or displacements) are all far better understood. These advancements have caused increases in seismic design forces from a factor of 1.5 in regions very near active faults (on the West Coast) to a factor of 2 to 3 in a few other areas of the country (e.g. Utah; Memphis, Tennessee). The damage to the first Olive View Hospital (Figure 8-2), in addition to other issues, was a result of inadequate zoning.

● Changes in Required Strength or Ductility

As discussed in Chapter 4, the required lateral strength of a seismic system is generally traded off with the **ductility** (the ability to deform inelastically—normally controlled by the type of detailing of the components and connections) of the system. Higher strength requires lower ductility and vice versa. The most significant changes in codes—reflecting better understanding of minimum requirements for life safety—are general increases in both strength and ductility. Many building types designed under previous seismic provisions, particularly in the 1950s, 1960s, and 1970s, are now considered deficient, including most concrete-moment frames and certain concrete shear walls, steel-braced frames, and concrete tilt-ups.

Other buildings designed with systems assumed to possess certain ductility have been proven inadequate. Figure 8-3 shows typical steel moment-frame damage in the Northridge earthquake caused by brittle behavior in a structural system previously thought to be of high ductility.

● Recognition of the Importance of Nonlinear Response

Historically, a limited amount of damage that absorbed energy and softened the building, thus attracting less force, was thought o reduce seismic response. Although this is still true, it is now recognized that the extent and pattern of damage must be controlled. Early codes required the design of buildings for forces three to six times less than the **elastic demand** (the forces that the building would see if there was no damage), assuming that the beneficial characteristics of damage would make up the difference. Unfortunately, buildings are not uniformly damaged, and the change in structural properties after damage (nonlinear response) often will concentrate seismic displacement in one location. For example, if the lower story of a building is much more flexible or weaker than the stories above, damage will concentrate at this level and act as a fuse, never allowing significant energy absorption from damage to the structure above. This concentrated damage can easily compromise the gravity load-carrying capacity of the structure at that level, causing collapse. Similarly, concrete shear walls were often "discontinued" at lower floors and supported on columns or beams. Although the supporting structure was adequately designed for code forces, the wall above is often much stronger than that and remains undamaged, causing concentrated and unacceptable damage in the supporting structure.

A final example of this issue can be seen by considering torsion. As explained in Chapter 4, Section 4.11, torsion in a building is a twisting in plane caused by an imbalance in the location of the mass and resisting elements. Older buildings were often designed with a concentration of lateral strength and stiffness on one end—an elevator/stair tower, for example—and a small wall or frame at the other end to prevent torsion. However, when the small element is initially damaged, its strength and stiffness changes, and the building as a whole may respond with severe torsion.

Current codes contain many rules to minimize configurations that could cause dangerous nonlinear response, as well as special design rules for elements potentially affected (e.g., columns supporting discontinuous shear walls). Olive View Hospital featured a weak first story in the main building, causing a permanent offset of more than one foot and near collapse; discontinuous shear walls in the towers caused a failure in the supporting beam and column frame, resulting in complete overturning of three of the four towers. Figure 8-4 shows a typical "tuck-under" apartment building in which the parking creates a weak story.

## 8.2.2 Philosophy Developed for Treatment of Existing Buildings

Building codes have long contained provisions to update life-safety features of buildings if the occupancy is significantly increased in number or level of hazard (transformation of a warehouse to office space, for example). As early as the mid-1960s, this concept started to be applied to seismic systems. Many older buildings contained entire structural systems no longer permitted in the code (e.g., URM, poorly reinforced concrete walls), and it quickly became obvious that 1) these components could not be removed, and 2) it was impractical and uneconomical to replace all older buildings. The "new" code could therefore not be applied directly to older buildings, and special criteria were needed to enable adaptive reuse while meeting the need to protect life safety of the occupants. In some cases, an entirely new and code-complying lateral system was installed, while leaving existing, now prohibited, construction in place. (This procedure was used in many school buildings in California after the Field Act was passed in 1933—up until the school seismic safety program was essentially completed in the 1960s.) This procedure proved very costly and disruptive to the building and was thought to discourage both improved seismic safety and general redevelopment.

Figure 8-2: Olive View Hospital. A brand new facility that was damaged beyond repair in the 1971 San Fernando earthquake due to a shaking intensity that exceeded what was expected at the site and a design that, although technically complying with code at the time, contained several structural characteristics now considered major deficiencies. The lateral system contained nonductile concrete frames, discontinuous shear walls, and a significant weak story.

Figure 8-3: An example of a recently designed building with seismic deficiencies not understood at the time of design. In this case, the deficiency was "pre-Northridge" moment-frame connections, which proved to be extremely brittle and unsatisfactory. Hundreds, if not thousands, of these buildings were designed and built in the two decades before the Northridge earthquake (1994).

Figure 8-4: A 1970s building type with a deficiency not prohibited by the code at the time—a tuck-under apartment with parking at the ground level, creating a weak story. Many of these buildings collapsed in the Northridge earthquake. Sixteen deaths occurred in the Northridge Meadows apartments, or 42% of those directly killed by the earthquake.

Figure 8-5: A tuck-under similar to Figure 8-4, but much more modern and designed at a time when the weakness of the parking level was more understood. In this case, the detailing of the small wood shear walls at that level was poor. This practice created another set of deficient "existing" buildings.

Figure 8-6: A tilt-up damaged in 1994. Despite suspicions that the code-required roof-to-wall ties were inadequate, it took several code cycles and incremental increases in requirements to obtain adequate code provision. Thousands of tilt-ups with inadequate connections exist in the West, although several jurisdictions are actively requiring retrofits.

SOURCE: LLOYD CLUFF

A philosophy quickly developed suggesting that existing buildings be treated differently from new buildings with regard to seismic requirements. First, archaic systems and materials would have to be recognized and incorporated into the expected seismic response, and secondly, due to cost and disruption, seismic design force levels could be smaller. The smaller force levels were rationalized as providing minimum life safety, but not the damage control of new buildings, a technically controversial and unproven concept, but popular. Commonly existing buildings were then designed to 75% of the values of new buildings—a factor that can still be found, either overtly or hidden, in many current codes and standards for existing buildings.

Occasionally, early standards for existing buildings incorporated a double standard, accepting a building that passed an evaluation using 75% of the code, but requiring retrofits to meet a higher standard, often 90% of the code.

## 8.2.3 Code Requirements Covering Existing Buildings

As the conceptual framework of evaluation and retrofit developed, legal and code requirements were also created. These policies and regulations can be described in three categories; active, passive, and post-earthquake. Active policies require that a defined set of buildings meet given seismic criteria in a certain time frame—without any triggering action by the owner. For example, all bearing-wall masonry buildings in the community must meet the local seismic safety criteria within ten years. Passive policies require minimum seismic standards in existing buildings only when the owner "triggers" compliance by some action—usually extensive remodeling, reconstruction, or addition. Post-earthquake policies developed by necessity after several damaging earthquakes, when it became obvious that repairing an obviously seismically poor building to its pre-earthquake condition was a waste of money. It then became necessary to develop triggers to determine when a building could simply be repaired and when it had to be repaired and retrofitted as well.

● Passive Code Provisions

As noted above, the development of requirements to seismically update a building under certain conditions mimicked economic and social policies well-established in building codes. Namely, the concept crystallized that if sufficient resources were spent to renew a building, particularly with a new occupancy, then the building should also be renewed seismically. Seismic "renewal" was defined as providing life safety, but not necessarily reaching the performance expected from a new building. A second kind of trigger—that could be termed "trigger of opportunity"—has also been used in some communities. These policies try to take advantage of certain conditions that make seismic improvements more palatable to an owner, such as retrofit of single-family dwellings at point of sale or requiring roof diaphragm upgrades at the time of re-roofing.

Triggers based on alterations to the building are by far the most common and will be discussed further here. These policies are somewhat logical and consistent with code practice, but they created two difficult socio-economic-technical issues that have never been universally resolved. The first is the definition of what level of building renewal or increase in occupancy-risk triggers seismic upgrading. The second is to establish the acceptable level of seismic upgrading.

Most typically, the triggering mechanisms for seismic upgrade are undefined in the code and left up to the local building official. The *Uniform Building Code*, the predecessor to the IBC in the western states, waffled on this issue for decades, alternately inserting various hard triggers (e.g., 50% of building value spent in remodels) and ambiguous wording that gave the local building official ultimate power. The use of this mechanism, whether well defined in local regulation or placed in the hands of the building official, ultimately reflects the local attitude concerning seismic safety. Aggressive communities develop easily and commonly triggered criteria, and passive or unaware communities require seismic upgrade only in cases of complete reconstruction or have poorly defined, easily negotiated triggers. For more specific information on seismic triggers in codes, see the accompanying sidebar.

When seismic improvement is triggered, the most common minimum requirement is life safety consistent with the overall code intent. However, the use of performance-based design concepts to establish equivalent technical criteria is a recent development and is not yet universally accepted. As indicated in the last section, the initial response to establishing minimum seismic criteria was to use the framework of the code provisions for new buildings with economic and technical adjustments as required. These adjustments included a lower lateral force level (a pragmatic response to the difficulties of retrofit), and special consideration for materials and systems not allowed by the provisions for new buildings.

EXISTING BUILDINGS–EVALUATION AND RETROFIT

## Box 1: Seismic Triggers in Codes

Events or actions that require owners to seismically retrofit their buildings are commonly called **triggers**. For example, in many communities, if an owner increases the occupancy risk (as measured by number of occupants, or by use of the building), they must perform many life-safety upgrades, including seismic ones. However, for practical and economic reasons, seldom does this trigger require conformance with seismic provision for new buildings, but rather with a special **life-safety** level of seismic protection, lower than that used for new buildings.

The code with the longest history in high-seismic regions, the *Uniform Building Code* (UBC), has long waffled on this issue. Besides the traditional code life-safety trigger based on clear-cut changes in occupancy, this code over the years has included provisions using hard triggers based on the cost of construction, and soft language that almost completely left the decision to the local building official. The last edition of this code, the 1997 UBC, basically allowed any (non-occupancy related) alteration as long as the seismic capacity was not made worse.

The codes and standards that will replace the UBC are based on a federally funded effort and published by FEMA as the NEHRP *Provisions*. These codes include the *International Building Code* (IBC), the National Fire Protection Agency (NFPA) and ASCE 7, a standard covering seismic design now ready for adoption by the other codes. This family of regulations has a common limit of a 5% reduction in seismic capacity before "full compliance" is required. This reduction could be caused by an increase in mass (as with an addition) or a decrease in strength (as with an alteration that places an opening in a shear wall). Full compliance in this case is defined as compliance with the provisions for new buildings that do not translate well to older buildings. It is unclear how this will be interpreted on the local level.

Many local jurisdictions, however, have adopted far more definitive triggers for seismic retrofit. San Francisco is a well-known example, perhaps because the triggers are fairly elaborately defined and because they have been in place for many years. In addition to the traditional occupancy-change trigger, San Francisco requires conformance with seismic

(continued over)

---

provisions specially defined for existing buildings when substantial nonstructural alterations are done on 2/3 or more of the number of stories within a two-year period, or when substantial structural alterations, cumulative since 1973, affect more than 30% of the floor and roof area or structure.

Although most jurisdictions leave this provision purposely loose, some also have adopted definitive triggers based on cost of construction, on the particular building type, and on various definitions of significant structural change.

Government, in some cases, has been much more aggressive in setting triggers to activate seismic retrofit, perhaps to create a lawful need for funds which otherwise would be difficult to obtain. The state of California has set a definitive list of seismic triggers for state-owned buildings: a) alteration cost exceeding 25% of replacement cost; b) change in occupancy; c) reduction of lateral load capacity by more than 5% in any story; d) earthquake damage reducing lateral load capacity by more than 10% at any story.

The federal government likewise, in RP 6, [NIST, 2002], also has definitive triggers: a) change in occupancy that increases the building's importance or level of use; b) alteration cost exceeding 50% of replacement cost; c) damage of any kind that has significantly degraded the lateral system; d) deemed to be a high seismic risk; and e) added to federal inventory though purchase or donation.

The regulations and policies governing any building, private or public, which will be significantly altered, should be researched in the planning stage to understand the effective seismic triggers, written or understood.

(Unreinforced masonry, for example, was not only prohibited as a structural system in zones of high seismicity, but also could not be used in a building at all.) Use of a lateral force level of 75% of that required for new buildings became fairly standard, but the treatment of archaic materials is highly variable from jurisdiction to jurisdiction.

Many local retrofit provisions are gradually being replaced by national guidelines and standards for seismic evaluation and retrofit (e.g. ASCE 31, 2003; FEMA 356, 2000, etc.). In addition, performance based seismic design is enabling a more direct approach to meeting a community's minimum performance standards—although this requires the policy-makers to decide what the minimum performance standard should be, a difficult task that crosses social, economic, and technical boundaries.

In summary, both the passive triggers for seismic retrofit and the design or performance criteria are often ill-defined and, at best, highly variable between jurisdictions. Design professionals should always determine the governing local, state, or federal regulations or policies when designing alterations or remodels on existing buildings.

● Active Code Provisions

Active code provisions result from policy decisions of a jurisdiction to reduce the community seismic risk by requiring seismic upgrading of certain buildings known to be particularly vulnerable to unacceptable damage. For the most part, these provisions are unfunded mandates, although low-interest loan programs have been developed in some cases. These risk reduction programs usually allow owners a lengthy period to perform the retrofit or to demolish the buildings—ten years or more. The standard for retrofit is also normally included in the law or regulation and is typically prescriptive, although performance-based design options are becoming more acceptable.

Two large-scale examples of active seismic code provisions were started by the state of California. The first was a program to reduce the risk from URM buildings. The state legislature, lacking the votes to simply require mitigation throughout the state, instead passed a law (SB 547-1986) that required local jurisdictions to develop inventories of these buildings in their area, to notify the owners that their building was considered hazardous, and to develop a community-wide hazard reduction plan.

Although not required to do so, most jurisdictions chose as their hazard reduction plan to pass active code ordinances giving owners of the buildings ten or so years to retrofit them. Over 10,000 URM buildings have been brought into compliance with these local ordinances, most by retrofit, but some by demolition (SSC, 2003).

The second program, created by SB 1953 in 1994 following the Northridge earthquake, gave California hospital owners until 2030 to upgrade or replace their hospitals to comply with state law governing new hospital buildings. The program's intention is to enable buildings to be functional following an earthquake. This law affected over 500 hospitals and over 2,000 buildings (Holmes, 2002). Although compliance is ongoing, this law has been problematic due to the high cost and disruption associated with retrofitting hospital buildings, and the highly variable economic condition of the health system as well as individual facilities.

Other examples include local ordinances to retrofit tilt-up buildings, less controversial because of the clear high vulnerability and low retrofit cost of these buildings. Similar to investigating local regulations regarding triggers, it is also wise to determine if any existing building planned for alterations is covered by (or will be covered in the foreseeable future) a requirement to retrofit. It is generally acknowledged that seismic improvements are easier to implement when done in association with other work on the building.

● Post-Earthquake Code Provisions

Following a damaging earthquake, many buildings may be closed pending determination of safety and necessary repairs. A lack of clear repair standards and criteria for re-occupancy has created controversy and denied owners use of their buildings after most damaging earthquakes. Assuming that the earthquake itself is the ultimate judge of seismic acceptability, many communities may take the opportunity in the post-earthquake period to require strengthening of buildings that are apparently seismically deficient due to their damage level. However, implementation of this theory incorporating conservative policies that require many retrofits may delay the economic recovery of the community. On the other hand, standards for repair and/or strengthening which are not conservative could lead to equal or worse damage in the next earthquake. It has also been observed that owners of historic, rent-controlled, or otherwise economically controlled buildings may have an incentive to demolish damaged buildings to the detriment of the community at large.

EXISTING BUILDINGS–EVALUATION AND RETROFIT

Traditionally, communities (building departments) have used color codes for several or all of the following categories of buildings following an earthquake:

**A.** Undamaged; no action required. If inspected at all, these buildings will be Green-tagged.

**B.** Damaged to a slight extent that will only require repair of the damage to the original condition. These buildings will generally be Green-tagged, but the category could also include some Yellow-tags.

**C.** Damaged to a greater extent that suggests such seismic weaknesses in the building that the overall building should be checked for compliance with minimum seismic standards. This will often require overall retrofit of the building. These buildings will generally be Red-tagged, but the category could also include some Yellow tags.

**C1.** (A subcategory of C). Damaged to an extent that the building creates a public risk that requires immediate mitigation, either temporary shoring or demolition. The ultimate disposition of these buildings may not be determined for several months. These buildings will all be Red-tagged.

The most significant categorization is the differentiation between B and C. The difference to an owner between being placed in one category or the other could be an expense on the order of 30%-50% of the value of the building, reflecting the added cost of retrofit to that of repair. Earthquakes being rare, few communities have been forced to create these policies, but a few have. Oakland, California, prior to the Loma Prieta earthquake, set a trigger based on the loss of capacity caused by the damage. If the damage was determined to have caused a loss of over 10% of lateral force capacity, then retrofit was triggered. Los Angeles and other southern California communities affected by the Northridge earthquake used a similar standard, but the 10% loss was applied to lines of seismic resistance rather than the building as a whole. These code regulations, although definitive, are problematic because of the technical difficulty of determining loss of capacity, particularly to the accuracy of 1% (a 1% change can trigger a retrofit).

The importance of this issue has been magnified by interpretation of federal laws that creates a tie between reimbursement of the cost of repair of certain local damage to the pre-existence and nature of these local damage triggers. Owners and designers of older existing buildings should be aware of such triggers that could affect them should they suffer damage from an earthquake. In some cases, it may be prudent for an owner to voluntarily retrofit a vulnerable building to avoid the possibility of being forced to do it in a post-earthquake environment as well as to possibly avoiding a long closure of the building.

## 8.3 THE FEMA PROGRAM TO REDUCE THE SEISMIC RISK FROM EXISTING BUILDINGS

In 1985, the Federal Emergency Management Agency (FEMA) recognized that the principal seismic risk in this country came from the existing building stock, the majority of which was designed without adequate seismic provisions. Following a national workshop that identified significant issues and potential educational and guideline projects that FEMA could lead, a program was launched that is still ongoing. In addition to providing education and technical guidelines in the area of high-risk existing buildings, other FEMA programs were also significant in enabling communities to understand and mitigate their seismic risk, most notably the development of the regional loss-estimating computer program, HAZUS. Most of these activities are documented as part of the FEMA "yellow book" series (so known because of its distinctive yellow covers), well known to engineers in this country and, in fact, around the world. Unfortunately, these documents are less known to architects, although many of them contain useful insights into not only the issues surrounding seismic evaluation and retrofit of existing buildings, but also into all aspects of seismic design.

### 8.3.1 FEMA-Sponsored Activity for Existing Buildings

Following is a summary of selected FEMA-sponsored projects beginning in the late 1980s. A full listing is given in FEMA 315, *Seismic Rehabilitation of Buildings, Strategic Plan 2005.*

● Rapid Visual Screening

FEMA 154: *Rapid Visual Screening of Buildings for Potential Seismic Hazards*, 1988, updated 2001

A method to enable an efficient first sorting of selected buildings into an adequately life-safe group and a second group that will require further evaluation. The evaluation was intended to be performed on the street in an hour or less per building. The first task is to assign the building to a predefined model building type and then identify additional characteristics that could refine the seismic vulnerability. The method has proven useful to efficiently generate an approximate mix of buildings that will properly characterize a community's vulnerability, but not to definitely rate individual buildings, due to the difficulty of identifying significant features from the street. Generally it is necessary to obtain access to the interior of a building, or, more commonly, it is even necessary to review drawings to confidently eliminate older buildings as potentially hazardous.

● Evaluation of Existing Buildings

FEMA 178: *NEHRP Handbook for Seismic Evaluation of Existing Buildings*, 1989 and FEMA 310: *Handbook for the Seismic Rehabilitation of Buildings: a Prestandard.*

A widely used guide to determine if individual buildings meet a nationally accepted level of seismic life safety. This method requires engineering calculations and is essentially prescriptive, which facilitates consistency and enables enforceability. This life-safety standard was adopted by, among others, the federal government and the state of California in certain programs. The prescriptive bar may, however, have been set too high, because very few older buildings pass. Since its original development, FEMA 178 has been refined and republished as FEMA 310, and finally was adopted as a Standard by the American Society of Civil Engineers as ASCE 31 in 2003.

● Techniques Used in Seismic Retrofit

FEMA 172: *NEHRP Handbook of Techniques for Seismic Rehabilitation of Existing Buildings*, 1992

In recognition of the lack of experience in seismic upgrading of buildings of most of the country's engineers and architects, this document outlined the basic methods of seismically strengthening a building, including conceptual details of typically added structural elements. The material recognizes the FEMA model building types, but is primarily organized around strengthening of structural components, perhaps making the material less directly accessible. The document also preceded by several years the publication of the analytical tools to design retrofits (FEMA 273, see below). For whatever reason, the publication went relatively unused, despite the fact that it contains useful information, particularly for architects unfamiliar with seismic issues.

● Financial Incentives

FEMA 198: *Financial Incentives for Seismic Rehabilitation of Hazardous Buildings*, 1990

To encourage voluntary seismic upgrading, this document described the financial incentives to do so, ranging from tax benefits to damage avoidance.

● Development of Benefit-Cost Model

FEMA 227: *A Benefit-Cost Model for the Seismic Rehabilitation of Buildings*, 1992

Due to the expected high cost of seismic rehabilitation, the need to provide a method to calculate the benefit-cost ratio of seismic retrofit was identified early in the program. This requires estimation of financial losses from earthquake damage resulting from a full range of ground-shaking intensity. Financial losses include direct damage to structural and nonstructural systems as well as business interruption costs. A controversial feature was the optional inclusion of the value of lives lost in the overall equation. The project was primarily to develop the model rather than to provide new research into expected damage or casualty rates. Thus, approximate relationships available at the time were used. However, the documentation concerning contributing factors to a benefit-cost analysis is quite complete, and a computerized functional spreadsheet version of the method was developed.

Although the use of benefit-cost analysis never became popular in the private sector, trials of the method indicated that very low retrofit costs, high business-interruption losses, or high exposure-to-casualty losses are required to result in a positive benefit-cost ratio. For example rehabilitation of tilt-up buildings is usually fairly inexpensive and usually prove cost effective. Similarly, buildings with high importance to a business or with high occupancy in areas of high seismicity also result in positive results. Despite the apparent overall results from this program, considerable rehabilitation activity continued, both in conditions expected to yield a positive benefit-cost and in other conditions (in many cases due to extreme importance given to life safety).

● Typical Costs of Seismic Rehabilitation

FEMA 156: *Typical Costs for Seismic Rehabilitation of Existing Buildings,* 1988, 2nd edition, 1994.

The second edition of this document collected case histories of constructed rehabilitation and completed reports judged to have realistic costs. A database was created to separate costs by primary influence factors: model building types, rehabilitation performance objectives, and seismicity. A serious difficulty in collection of accurate data was the inevitable mixing of pure rehabilitation costs and associated costs such as life-safety upgrades, the American Disabilities Act (ADA), and even remodels. Although a large amount of data was collected, there was not nearly enough to populate all combinations of the factors. Nevertheless, a method was developed to use the data to make estimates of costs for given situations. The major problem was that the coefficient of variation of rehabilitation costs, for any given situation, is very high due to high variability in the extent of seismic deficiencies. The information collected is probably most useful to estimate costs for large numbers of similar buildings where variations will average out. Use of the method to accurately estimate the cost of a single building is not recommended, although even the ranges given could be useful for architects and engineers not familiar with retrofit issues.

● Technical Guidelines for Seismic Rehabilitation

FEMA 273: *NEHRP Guidelines for the Seismic Rehabilitation of Buildings,* 1997 and FEMA 356 *Prestandard and Commentary for the Seismic Rehabilitation of Buildings.*

This document, developed over five years by over 70 experts, was the culmination of the original program. Previously, the most common complaint from engineers and building officials was the lack of criteria for seismic retrofit. FEMA 273 incorporated performance-based engineering, state-of-the-art nonlinear analysis techniques, and an extensive commentary to make a significant contribution to earthquake engineering and to focus laboratory research on development of missing data. The document broke away from traditional code methods and in doing so, faced problems of inconsistency with the design of new buildings. Improvements were made in a follow-up document, FEMA 356, but the practical results from use of the method indicate that considerable judgment is needed in application. Work is continuing to improve analysis methods and in methods to predict the damage level to both components of a building and to the building as a whole. Even given these difficulties, the document has become the standard of the industry.

- ● Development of a Standardized Regional Loss Estimation Methodology—HAZUS

National Institute of Building Sciences, *Earthquake Loss Estimation Methodology, Technical Manual* (for latest release).

A good summary of this development is contained in a paper by Whitman, et al. in EERI *Earthquake Spectra*, vol 13, no 4. FEMA also maintains a HAZUS website, www.fema.gov/hazus.

The development of a standardized regional loss estimation methodology was not in the original FEMA plan to reduce risks in existing buildings. However, this development has had a major impact in educating local officials about their seismic risks and estimating the level of risk around the country in a standard and comparable way.

In 1990, when the development of HAZUS began, the primary goals were to raise awareness of potential local earthquake risks, to provide local emergency responders with reasonable descriptions of post-earthquake conditions for planning purposes, and to provide consistently created loss estimates in various regions to allow valid comparison and analysis. The loss estimation methodology was intended to be comprehensive and cover not only building losses, but also damage to transportation systems, ports, utilities, and critical facilities. A technically defensible methodology was the goal, not necessarily an all-encompassing software package. When it became obvious that the methodology was far more useful and could be more consistently applied as software, HAZUS was born. The program uses census data and other available physical and economic databases to develop, on a first level of accuracy, a model of local conditions. Expected or speculated seismic events can be run and losses estimated. Losses include direct damage, business interruption, and casualties, as well as loss of utilities, loss of housing units, and many other parameters of use to emergency planners. The building inventory uses the FEMA model building types and an analysis method closely tied to FEMA 273, linking HAZUS to other FEMA-sponsored work regarding existing buildings.

Subsequent to the original development activity, HAZUS was expanded to create loss estimates for wind and flood.

● Incremental Rehabilitation

FEMA 395, *Incremental Seismic Rehabilitation of School Buildings (K-12),* 2003.

This is the first in a series of manuals that FEMA (U.S. Department of Homeland Security) intends to develop for various occupancy types including, for example, schools, hospitals, and office buildings. The concept is based on the fact that seismic strengthening activities are more efficiently accomplished in conjunction with other work on the building, and such opportunities should be identified and exploited even if only part of a complete rehabilitation is accomplished. This is perhaps most applicable to K-12 school buildings because of their relatively small size and ongoing maintenance programs. FEMA model building types are again used to categorize potential opportunities in different conditions. As is pointed out in the manual, this technique has to be applied with care to avoid an intermediate structural condition that is worse than the original.

## 8.3.2 The FEMA Model Building Types

Most of these developments were part of the integrated plan developed in 1985. As such, FEMA coordinated the projects and required common terminology and cross-references.

The most successful and virtually standard-setting effort was the creation of a set of model building types to be used for the characterization of existing buildings. The model building types are based primarily on structural systems rather than occupancy, but have proven extremely useful in the overall program. Model building types are defined by a combination of the gravity-load carrying system and the lateral-load carrying system of the building. Not every building type ever built in the country and certainly not the world is represented, but the significant ones are, and the relative risks of a community can well be represented by separating the local inventory into these types. Of course, there was no attempt to represent every "modern" building type because they are not considered hazardous buildings. However, with minor sub-categorization that has occurred with successive documents, the majority of buildings, new or old, now can be assigned a model building type. The test of the usefulness came with the successful development of HAZUS using the model building type because this program needed a reasonably simple method to characterize the seismic vulnerability of inventories of buildings across the country.

Currently, no single FEMA document contains a graphic and clear description of the model building types, although engineers can generally determine the correct category. Because of the ubiquitous FEMA-developed documents, guidelines, and standards regarding existing buildings, and their common use by engineers, such descriptions are included here to facilitate communication with architects. The types are illustrated on pages 8-23 through 8-31. Table 8-3, at the end of the chapter, presents a summary of the performance charactoristics and commom rehabilitation techniques.

## 8.4 SEISMIC EVALUATION OF EXISTING BUILDINGS

Not all older buildings are seismically at risk. If they were, the damage from several earthquakes in this country, including the 1971 San Fernando and the 1984 Northridge events, would have been devastating, because much of the inventory affected was twenty or more years old. Often, strong ground shaking from earthquakes significantly damages building types and configurations well known to be vulnerable, and occasionally highlights vulnerabilities previously unrealized. For example, the Northridge earthquake caused damage to many wood-frame buildings—mostly apartments—and relatively modern steel moment-frame buildings, both previously considered to be of low vulnerability. It is natural to catalogue damage after an earthquake by buildings with common characteristics, the most obvious characteristic being the construction type, and the secondary characteristic being the configuration. Both of these parameters are central to processes developed to identify buildings especially vulnerable to damage before the earthquake. In fact, the categorization of damage by building type is primarily what led to the development of the FEMA Model Building Types discussed in Section 8.3.2.

However, only in the most vulnerable building types does damage occur relatively consistently. For example, at higher levels of shaking, the exterior walls of unreinforced masonry bearing-wall buildings have relatively consistently fallen away from their buildings in many earthquakes, ever since this building type was built in large numbers in the late 19th century. More recently, a high percentage of "pre-Northridge" steel

## FEMA Building Type W1     WOOD LIGHT FRAME  (small residence)

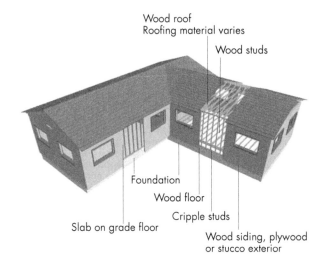

Wood roof
Roofing material varies

Wood studs

Foundation

Wood floor

Cripple studs

Slab on grade floor

Wood siding, plywood
or stucco exterior

These buildings are generally single-family dwellings of one and two stories. Floor and roof framing consists of wood joists or rafters supported on wood studs spaced no more than 24 inches apart.  The first floor may be slab on grade or wood raised above grade with cripple stud walls and post-and-beam supports. Lateral support is provided with shear walls of plywood, stucco, gypsum board, and a variety of other materials. Most often there is no engineering design for lateral forces.

## FEMA Building Type W1A     WOOD LIGHT FRAME (multi-unit residence)

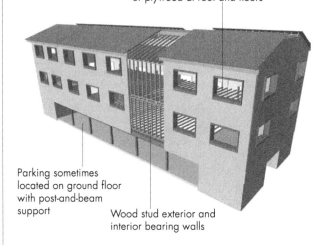

Wood joist floors with sheathing
or plywood at roof and floors

Parking sometimes located on ground floor with post-and-beam support

Wood stud exterior and interior bearing walls

These buildings are framed with the same systems as W1 buildings but are most often multiple-story large residential-type structures, and, unless very old, are engineered. A common seismic deficiency is the tuck-under parking at the ground story that creates a soft or weak story. This building type is also often built on top of a one story concrete parking structure.

## FEMA Building Type W2    WOOD FRAME  (commercial and industrial)

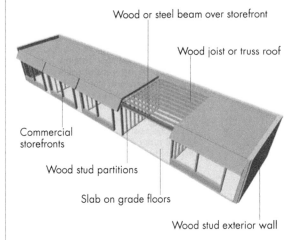

Wood or steel beam over storefront

Wood joist or truss roof

Commercial storefronts

Wood stud partitions

Slab on grade floors

Wood stud exterior wall

These buildings are commonly commercial or smaller industrial buildings and are constructed primarily of wood framing. The floor and roof framing consists of wood joists and wood or steel trusses, glulam or steel beams, and wood posts or steel columns. Lateral forces are resisted by wood diaphragms and exterior stud walls sheathed with plywood, stucco, or wood sheathing, or sometimes rod bracing or a spot steel-braced frame. Large wall openings are common for storefronts or garage openings.  This building type is also often used for schools, churchs and clubhouses.

## FEMA Building Type S1    STEEL MOMENT FRAMES

Vertical shafts of nonstructural materials

Steel beams and columns

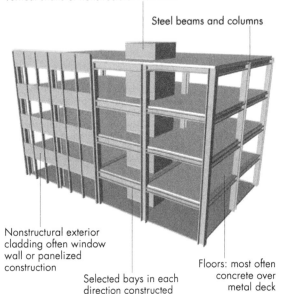

Nonstructural exterior cladding often window wall or panelized construction

Selected bays in each direction constructed as moment frames. See chapter 3.

Floors: most often concrete over metal deck

These buildings consist of an essentially complete frame assembly of steel beams and columns. Lateral forces are resisted by moment frames that develop stiffness through rigid connections of the beam and column created by angles, plates and bolts, or by welding. Moment frames may be developed on all framing lines or only in selected bays. It is significant that no structural walls are required. Floors are cast-in-place concrete slabs or metal deck and concrete. This building is used for a wide variety of occupancies such as offices, hospitals, laboratories, and academic and government buildings.

The S1A building type is similar but has floors and roof that act as flexible diaphragms, such as wood or uptopped metal deck. One family of these buildings are older warehouse or industrial buildings, while another more recent use is for small office or commercial buildings in which the fire rating of concrete floors is not needed.

## FEMA Building Type S2  STEEL-BRACED FRAMES

Braced frames often placed within shaft walls

Steel beams and columns

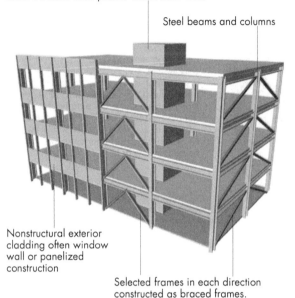

Nonstructural exterior cladding often window wall or panelized construction

Selected frames in each direction constructed as braced frames. See chapter 3.

These buildings consist of a frame assembly of steel columns and beams. Lateral forces are resisted by diagonal steel members placed in selected bays. Floors are cast-in-place concrete slabs or metal deck and concrete. These buildings are typically used for buildings similar tos teel-moment frames, although are more often low rise.

The S2A building type is similar but has floors and roof that act as flexible diaphragms such as wood, or uptopped metal deck. This is a relatively uncommon building type and is used mostly for smaller office or commercial buildings in which the fire rating of concrete floor is not needed.

## FEMA Building Type S3    STEEL LIGHT FRAMES

Light-gauge metal cladding

Steel bents in short direction

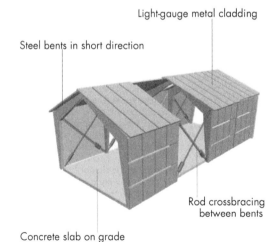

Rod crossbracing between bents

Concrete slab on grade

These buildings are one story, pre-engineered and partially prefabricated, and normally consist of transverse steel bents and light purlins. The roof and walls consist of lightweight metal, fiberglass, or cementitious panels. Lateral forces are resisted by the transverse steel bents acting as moment frames, and light rod diagonal bracing in the longitudinal direction. The roof diaphragm is either metal deck or diagonal rod bracing. These buildings are mostly used for industrial or agricultural occupancies.

## FEMA Building Type S4   STEEL FRAMES  with concrete shearwalls

"Punched" concrete exterior walls
are an alternate shear-wall configuration

Vertical shafts often constructed of concrete

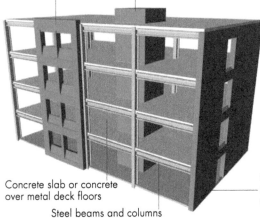

Concrete slab or concrete
over metal deck floors

Steel beams and columns

Concrete walls placed in
selected interior and and
exterior bays in each direction

These buildings consist of an essentially complete frame assembly of steel beams and steel columns. The floors are concrete slabs or concrete fill over metal deck. The buildings feature a significant number of concrete walls effectively acting as shear walls, either as vertical transportation cores, isolated in selected bays, or as a perimeter wall system. The steel column-and-beam system may act only to carry gravity loads or may have rigid connections to act as a moment frame. This building type is generally used as an alternate for steel moment or braced frames in similar circumstances. These buildings will usually be mid- or low-rise.

## FEMA Building Type S5    STEEL FRAMES with infill masonry walls

Interior partitions or shaft walls
often built with clay tile

Steel beams and columns

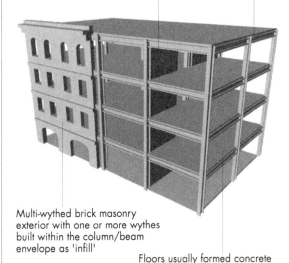

Multi-wythed brick masonry
exterior with one or more wythes
built within the column/beam
envelope as 'infill'

Floors usually formed concrete

This is normally an older building that consists of an essentially complete frame assembly of steel floor beams or trusses and steel columns. The floor consists of masonry flat arches, concrete slabs or metal deck, and concrete fill. Exterior walls and possibly some interior walls, are constructed of unreinforced solid clay brick, concrete block, or hollow-clay tile masonry infilling the space between columns and beams. Windows and doors may be present in the infill walls, but to act effectively as shear-resisting elements, the infill masonry must be constructed tightly against the columns and beams. Although relatively modern buildings in moderate or low seismic regions are built with unreinforced masonry exterior infill walls, the walls are generally not built tight against the beams and columns and therefore do not provide shear resistance. The buildings intended to fall into this category feature exposed clay brick masonry on the exterior and are common in commercial areas of cities with occupancies of retail stores, small offices, and hotels.

The S5A building type is similar but has floors and roof that act as flexible diaphragms, such as wood or uptopped metal deck. These buildings will almost all date to the 1930s and earlier, and were originally warehouses or industrial buildings.

## FEMA Building Type C1 CONCRETE MOMENT FRAMES

Vertical shafts of nonstructural materials

Concrete beams and columns

Nonstructural exterior cladding, often window wall or panelized construction

Selected bays in each direction constructed as moment frames

Floors: most often formed or precast concrete

These buildings consist of concrete framing, either a complete system of beams and columns or columns supporting slabs without gravity beams. Lateral forces are resisted by moment frames that develop stiffness through rigid connections of the column and beams placed in a given bay. Moment frames may be developed on all framing lines or only in selected bays. It is significant that no structural walls are required. Floors are cast-in-place or precast concrete. Buildings with concrete moment frames could be used for most occupancies listed for steel moment frames, but are also used for multistory residential buildings.

The C1A building type is similar but has floors and roof that act as flexible diaphragms, such as wood or uptopped metal deck. This is a relatively unusual building type, but might be found as older warehouse-type buildings or small office occupancies.

## FEMA Building Type C3 CONCRETE FRAMES with infill masonry shear walls

Interior partitions or shaft walls often built with clay tile

Concrete beams and columns or slabs and columns

Multi-wythed brick masonry exterior, one or more wythes built within the column/beam envelope as "infill"

Floors usually formed concrete

These buildings consist of concrete framing, either a complete system of beams and columns or columns supporting slabs without gravity beams. Exterior walls and possibly some interior walls are constructed of unreinforced solid clay brick, concrete block, or hollow clay tile masonry infilling the space between columns and beams. Windows and doors may be present in the infill walls, but to act effectively as shear-resisting elements, the infill masonry must be constructed tightly against the columns and beams. The building type is similar to S5, but is more often used for industrial and warehouse occupancies.

The C3A building type is similar but has floors and roof that act as flexible diaphragms, such as wood, or uptopped metal deck. Thisbuilding type not often found except as one-story industrial buildings.

# FEMA Building Type C2 CONCRETE SHEAR WALLS

## with bearing walls

Precast or formed floors span
between bearing walls

Concrete interior bearing walls

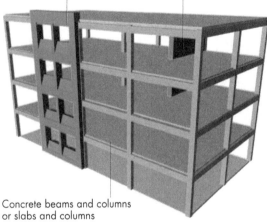

Concrete exterior wall

## with gravity frames

Exterior walls: punched concrete shearwalls
or concrete pier-and-spandrel system

Selected interior walls may be
concrete shear walls

Concrete beams and columns
or slabs and columns

Concrete shear walls are concrete walls
in a building design to provide lateral
stiffness and strength for lateral loads.
There are two main types of shear-wall
buildings, those in which the shear
walls also carry the gravity loads (with
bearing walls), and those in which a
column-supported framing system
carries the gravity loads (with gravity
frame).

In the **bearing wall** type, all walls
usually act as both bearing and shear
walls. The building type is similar and
often used in the same occupancies as
type RM2, namely in mid- and low-rise
hotels and motels. This building type is
also used in residential apartment/
condo-type buildings.

In **gravity frame** buildings, shear
walls are either strategically placed
around the plan, or at the perimeter.
Shear-wall systems placed around the
entire perimeter must contain the
windows, and other perimeter
openings are called punched shear
walls. These buildings were commonly
built in the 1950s and 1960s for a
wide variety of most institutional
occupancy types.

The C2A building type is similar, but
has floors and roof that act as flexible
diaphragms such as wood, or
uptopped metal deck. C2A buildings
are normally bearing-wall buildings.
These buildings are similar to
building-type RM1 and are used for
similar occupancies- such as small
office or commercial and sometimes
residential.

## FEMA Building Type PC1    TILT-UP CONCRETE shear walls

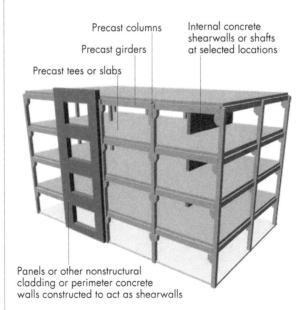

Plywood roof

Wood joists

Wood purlins

Steel or glulam girders

Roof supported on exterior panels, cast-in-place concrete columns, or independant steel columns

Precast exterior wall panels

These buildings are constructed with perimeter concrete walls precast on the site and tilted up to form the exterior of the buildings, to support all or a portion of the perimeter roof load, and to provide seismic shear resistance. These buildings are commonly one-story with a wood joist and plywood roof or sometimes with a roof of steel joists and metal deck. Two-story tilt-ups usually have a steel-framed second floor with metal deck and concrete and a wood roof. Tilt-up walls that support roof load are very common on the West Coast; due to economical construction cost, they are used for many occupancies, including warehouses, retail stores, and offices. In other parts of the country, these buildings more often have an independent load-carrying system on the inside face of the walls.

The PC1A building is similar but features all floors and/or roof constructed of materials that form a rigid diaphragm, normally concrete. This building type is similar to PC2.

## FEMA Building Type PC2    PRECAST CONCRETE FRAMES with shear walls

Precast columns

Precast girders

Precast tees or slabs

Internal concrete shearwalls or shafts at selected locations

Panels or other nonstructural cladding or perimeter concrete walls constructed to act as shearwalls

These buildings consist of concrete columns, girders, beams and/or slabs that are precast off the site and erected to form a complete gravity-load system. Type PC2 has a lateral force-resisting system of concrete shear walls, usually cast-in-place. Many garages have been built with this system The building type is most common in moderate and low seismic zones and could be used for many different occupancies in those areas.

The PC2A building is similar but obtains lateral support from specially connected precast girders and columns that form moment frames. Until recently, precast moment frames have not been allowed in regions of high seismicity, and these buildings will essentially only be found in moderate or low seismic zones.

**FEMA Building Type RM1   REINFORCED MASONRY WALLS with flexible diaphrams**

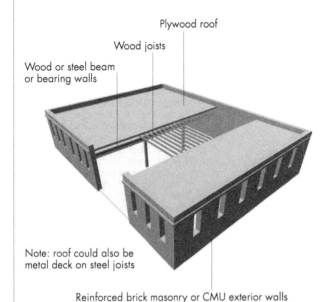

Plywood roof

Wood joists

Wood or steel beam or bearing walls

Note: roof could also be metal deck on steel joists

Reinforced brick masonry or CMU exterior walls

These buildings take a variety of configurations, but they are characterized by reinforced masonry walls (brick cavity wall or CMU) with flexible diaphragms, such as wood or metal deck. The walls are commonly bearing, but the gravity system often also contains post-and-beam construction of wood or steel. Older buildings of this type are generally small and were used for a wide variety of occupancies and are configured to suit. Recently, the building type is commonly used for one-story warehouse-type occupancies similar to tilt-up buildings.

---

**FEMA Building Type RM2   REINFORCED MASONRY WALLS with stiff diaphrams**

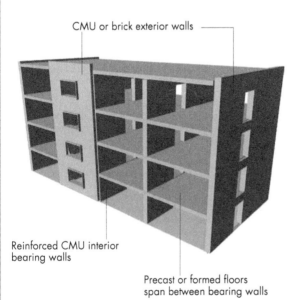

CMU or brick exterior walls

Reinforced CMU interior bearing walls

Precast or formed floors span between bearing walls

This building consists of reinforced masonry walls and concrete slab floors that may be either cast-in-place or precast. This building type is often used for hotel and motels and is similar to the concrete bearing-wall type C2.

---

EXISTING BUILDINGS–EVALUATION AND RETROFIT

2-4 wythe brick masonry exterior bearing walls

Wood joists or trusses with wood sheathing

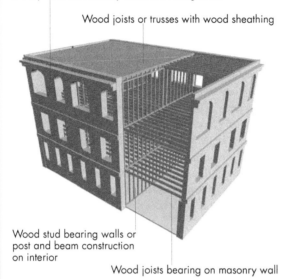

Wood stud bearing walls or post and beam construction on interior

Wood joists bearing on masonry wall

This building consists of unreinforced masonry bearing walls, usually at the perimeter and usually brick masonry. The floors are wood joists and wood sheathing supported on the walls and on interior post-and-beam construction or wood-stud bearing walls. This building type is ubiquitous in the U.S. and was built for a wide variety of uses, from one-story commercial or industrial occupancies, to multistory warehouses, to mid-rise hotels. Unfortunately, it has consistently performed poorly in earthquakes. The most common failure is an outward collapse of the exterior walls, caused by loss of lateral support due to separation of the walls from the floor/roof diaphragm.

The URMA building is similar. but features all floors and/or roof constructed of materials that form a rigid diaphragm, usually concrete slabs or steel joists with flat-arched unreinforced masonry.

moment-frame buildings have received damage to their beam-column connections when subjected to strong shaking. Even in these cases, the damage is not 100% consistent and certainly not 100% predictable. In building types with less vulnerability, the damage has an even higher coefficient of variation. Engineers and policymakers, therefore, have struggled with methods to reliably evaluate existing buildings for their seismic vulnerability.

As discussed in Section 8.2, the initial engineering response was to judge older buildings by their capacity to meet the code for new buildings, but it became quickly apparent that this method was overly conservative, because almost every building older than one or two code-change cycles would not comply—and thus be considered deficient. Even when lower lateral force levels were used, and the presence of archaic material was not, in itself, considered a deficiency, many more buildings were found deficient than was evidenced in serious earthquake damage. Thus, policymakers have generally been successful in passing active retrofit provisions (see Section 8.2.3) only in the most vulnerable buildings, such as URM and tilt-ups, where damage has been significant and consistent, and individual building evaluation is not particularly significant.

The evaluation of existing buildings typically starts with identification of the building type and damaging characteristics of configuration (e.g., soft story). This can be done rapidly and inexpensively but, except for a few vulnerable building types, is unreliable when taken to the individual building level. Engineers and code writers have also developed intermediate levels of evaluation in which more characteristics are identified and evaluated, many by calculation. In the last decade, more sophisticated methods of analysis and evaluation have been developed that consider the nonlinear response of most structures to earthquakes and very detailed material and configuration properties that will vary from building to building.

## 8.4.1 Expected Performance by Building Type

As previously mentioned, damage levels after earthquakes are collected and generally assigned to bins of common characteristics, most commonly the level of shaking, building material and type, and configuration. Combined with numerical lateral-force analysis of prototype buildings, this information can be analyzed statistically. The three primary parameters - building type, shaking level, and damage level - are often displayed together in a damage probability matrix similar to Table 8-1. The variability of damage is such that for any shaking level, as shown in the columns, there is normally a probability that some buildings will be in each damage state. The probabilities in these tables can be interpreted as the percentage of a large number of buildings expected to be in each damage state, or the chances, given the shaking level, that an individual building of this type with be damaged to each level.

Statistical information such as this is used in several ways:

● Identification of clearly vulnerable or dangerous buildings to help establish policies of mitigation

Many extremely vulnerable building types or components can be identified by observation without statistical analysis, including URM, soft-story "tuck-under" apartment buildings, the roof-to-wall connection in tilt-up

Table 8-1: Typical Form of Damage Probability Matrix

| Damage Level | Strength of Ground Motion (Peak Ground Acceleration, Spectral Acceleration, or Modified Mercalli Intensity [MMI].  MMI shown here) | | | | |
|---|---|---|---|---|---|
| | VI | VII | VIII | IX | X |
| None | .84 | .65 | .05 | .02 | - |
| Slight | .15 | .28 | .75 | .31 | .20 |
| Moderate | .01 | .03 | .10 | .47 | .50 |
| Severe | - | .03 | .08 | .15 | .20 |
| Complete | - | .01 | .02 | .05 | .10 |

buildings, residences with cripple wall first-floor construction, and connections of pre-Northridge steel moment-frames.  The clearly and more consistently dangerous building types have often generated enough community concern to cause the creation of policies to mitigate the risks with retrofit.  For a combination of reasons, URMs and tilt-ups currently are the targets of the most active mitigation policies.

● Earthquake Loss Estimation

Regional earthquake loss estimates have been performed for forty or more years to raise awareness in the community about the risks from earthquakes and to facilitate emergency planning.  Given an approximate distribution of the building inventory and a map of estimated ground motion from a given earthquake, damage-probability matrices (or similar data) can be used to estimate damage levels to the building stock.  From the damage levels, economic loss, potential casualties, and business interruption in a community can be estimated.

Starting in 1991, FEMA began a major program to develop a standard way of performing such loss estimations to facilitate comparative loss estimates in various parts of the country.  This program resulted in a computer program, HAZUS, described briefly in Section 8.3.1.

● Formal Economic Loss Evaluations (e.g. Probable Maximum Loss or PML)

Since consensus loss relationships became available (ATC, 1985), a demand has grown to include an estimate of seismic loss in "due-diligence" studies done for purchase of buildings, for obtaining loans for purchase or refinance, or for insurance purposes. An economic loss parameter, called Probable Maximum Loss, has become the standard measuring stick for these purposes. The PML for a building is the pessimistic loss (the loss suffered by the worst 10% of similar buildings) for the worst shaking expected at the site (which gradually became defined as the shaking with a 500-year return period, similar to the code design event). Although a detailed analysis can be performed to obtain a PML, most are established by building type and a few observable building characteristics. Because of the high variability in damage and the relatively incomplete statistics available, PMLs are not very reliable, particularly for an individual building.

● Rapid Evaluation

As foreseen by FEMA's original plan for the mitigation of risks from existing buildings, a rapid evaluation technique should be available to quickly sort the buildings into three categories: obviously hazardous, obviously acceptable, and uncertain. The intent was to spend less than two hours per building for this rapid evaluation. Under the plan, the uncertain group would then be evaluated by more detailed methods. The results of FEMA's development efforts, FEMA 154 (Section 8.3.1) is fairly sophisticated but, because of the large amount of unknown building data that is inherent in the system, for an individual building, is unreliable. The sorting method is probably quite good for estimating the overall vulnerability of a community because of the averaging effect when estimating the risk of many buildings.

## 8.4.2 Evaluation of Individual Buildings

Engineers have been seismically evaluating existing buildings for many years, whether by comparing the conditions with those required by the code for new buildings, by using some local or building-specific standard (e.g. URMs), or by using their own judgment. These methods are still used, as well as very sophisticated proprietary methods developed within private offices, but the majority of evaluations are now tied in some way to the general procedures of ASCE 31-03, *Seismic Evaluation of Existing Buildings* (ASCE, 2003), that in 2003 became a national standard. There are three levels of evaluation in the standard called tiers, which, not accidentally, are similar to the standard of practice prior to the standardization process. These levels of evaluation are briefly described below, as well as similar methods that fall in the same categories.

However, before beginning a seismic evaluation, particularly of a group of buildings, it is logical to assume that buildings built to modern codes must meet some acceptable standard of life safety. Due to the large number of older buildings, the effort to eliminate some from consideration resulted in several well-known milestone years. First it was compliance with the 1973 *Uniform Building Code* (UBC) or equivalent. After study and reconsideration of relatively major changes made in the 1976 UBC, this code was used as a milestone. Primarily caused by life-threatening damage to various building types in the 1994 Northridge earthquake and subsequent code changes, a relatively complex set of milestone years was developed. ASCE 31 contains such a set of code milestone years for each of the several codes used in this country over the last thirty years. Although ASCE 31 suggests that compliance with these codes is only a recommended cut-off to not require evaluations for life safety, in all but very unusual situations, the table can be accepted. This table, Table 3-1 of ASCE 3,1 is reproduced as Figure 8-7.

● Initial Evaluation (ASCE 31 Tier 1)

The ASCE Tier 1 evaluation is similar to FEMA's Rapid Evaluation in that it is based on the model building type and certain characteristics of the building. The significant difference is that structural drawings, or data equivalent to structural drawings, are required to complete the evaluation, and the evaluation will take several days rather than several hours. After identifying the appropriate FEMA Building Type, a series of prescriptive requirements are investigated, most of which do not require calculations.

If the building is found to be noncompliant with any requirement, it is potentially seismically deficient. After completing the investigation of a rather exhaustive set of requirements, the engineer reviews the list of requirements with which the building does not comply, and decides if the building should be categorized as noncompliant or deficient. A conservative interpretation of the method is that any single noncompliance is sufficient to fail the building, but most engineers exercise their judgement in cases of noncompliance with only a few requirements. Historically, this method has developed with the pass/fail criterion of life safety, but the final ASCE Standard includes criteria for both Life Safety and Immediate Occupancy, a performance more closely related to continued use of the building. Because of the importance associated with the Immediate Occupancy performance level, a building cannot pass these requirements with only a Tier 1 analysis.

---

EXISTING BUILDINGS–EVALUATION AND RETROFIT

## Table 3-1. Benchmark Buildings

| Building Type[1,2] | Model Building Seismic Design Provisions | | | | | FEMA 178[ls] | FEMA 310[ls,lo] | CBC[lo] |
|---|---|---|---|---|---|---|---|---|
| | NBC[ls] | SBC[ls] | UBC[ls] | IBC[ls] | NEHRP[ls] | | | |
| Wood Frame, Wood Shear Panels (Type W1 & W2) | 1993 | 1994 | 1976 | 2000 | 1985 | * | 1998 | 1973 |
| Wood Frame, Wood Shear Panels (Type W1A) | * | * | 1997 | 2000 | 1997 | * | 1998 | 1973 |
| Steel Moment Resisting Frame (Type S1 & S1A) | * | * | 1994[4] | 2000 | ** | * | 1998 | 1995 |
| Steel Braced Frame (Type S2 & S2A) | 1993 | 1994 | 1988 | 2000 | 1991 | 1992 | 1998 | 1973 |
| Light Metal Frame (Type S3) | * | * | * | 2000 | * | 1992 | 1998 | 1973 |
| Steel Frame w/ Concrete Shear Walls (Type S4) | 1993 | 1994 | 1976 | 2000 | 1985 | 1992 | 1998 | 1973 |
| Reinforced Concrete Moment Resisting Frame (Type C1)[3] | 1993 | 1994 | 1976 | 2000 | 1985 | * | 1998 | 1973 |
| Reinforced Concrete Shear Walls (Type C2 & C2A) | 1993 | 1994 | 1976 | 2000 | 1985 | * | 1998 | 1973 |
| Steel Frame with URM Infill (Type S5, S5A) | * | * | * | 2000 | * | * | 1998 | * |
| Concrete Frame with URM Infill (Type C3 & C3A) | * | * | * | 2000 | * | * | 1998 | * |
| Tilt-up Concrete (Type PC1 & PC1A) | * | * | 1997 | 2000 | * | * | 1998 | * |
| Precast Concrete Frame (Type PC2 & PC2A) | * | * | * | 2000 | * | 1992 | 1998 | 1973 |
| Reinforced Masonry (Type RM1) | * | * | 1997 | 2000 | * | * | 1998 | * |
| Reinforced Masonry (Type RM2) | 1993 | 1994 | 1976 | 2000 | 1985 | * | 1998 | * |
| Unreinforced Masonry (Type URM)[5] | * | * | 1991[6] | 2000 | * | 1992 | * | * |
| Unreinforced Masonry (Type URMA) | * | * | * | 2000 | * | * | 1998 | * |

[1] Building Type refers to one of the Common Building Types defined in Table 2-2.
[2] Buildings on hillside sites shall not be considered Benchmark Buildings.
[3] Flat Slab Buildings shall not be considered Benchmark Buildings.
[4] Steel Moment-Resisting Frames shall comply with the 1994 UBC Emergency Provisions, published September/October 1994, or subsequent requirements.
[5] URM buildings evaluated using the ABK Methodology (ABK, 1984) may be considered benchmark buildings.
[6] Refers to the GSREB or its predecessor, the UCBC (Uniform Code of Building Conservation).

[ls] Only buildings designed and constructed or evaluated in accordance with these documents and being evaluated to the Life-Safety Performance Level may be considered Benchmark Buildings.
[lo] Buildings designed and constructed or evaluated in accordance with these documents and being evaluated to either the Life-Safety or Immediate Occupancy Performance Level may be considered Benchmark Buildings.

\* No benchmark year; buildings shall be evaluated using this standard.
\*\* Local provisions shall be compared with the UBC.

NBC—Building Officials and Code Administrators, *National Building Code.*
SBC—Southern Building Code Congress, *Standard Building Code.*
UBC—International Conference of Building Officials, *Uniform Building Code.*
GSREB—International Conference of Building Officials, *Guidelines for Seismic Retrofit of Existing Buildings.*
IBC—International Code Council, *International Building Code.*
NEHRP—Federal Emergency Management Agency, *NEHRP Recommended Provisions for the Development of Seismic Regulations for New Buildings.*
CBC—California Building Standards Commission, *California Building Code, California Code of Regulations, Title 24.*

Figure 8-7: From ASCE 31-03, showing building types that might be considered "safe" due to the code under which they were designed.

SOURCE: ASCE 31-01, *SEISMIC EVALUATION OF EXISTING BUILDINGS,* THE STRUCTURAL ENGINEERING INSTITUTE OF THE AMERICAN SOCIETY OF CIVIL ENGINEERS, 2003 (REPRINTED WITH PERMISSION OF ASCE).

● Intermediate Evaluation (ASCE 31 Tier 2)

The ASCE intermediate level of evaluation, called Tier 2, is similar in level of effort of historical nonstandardized methods. Normally, an analysis of the whole building is performed and the equivalents of stress checks are made on important lateral force-resisting components. This analysis is done in the context and organization of the set of requirements used in Tier 1, but the process is not unlike seismic analysis traditionally performed for both evaluation and design of new buildings. ASCE 31 includes the requirements for both the LifeSafety and Immediate Occupancy performance levels for a Tier 2 Evaluation.

● Detailed Evaluation (ASCE 31 Tier 3)

The most detailed evaluations are somewhat undefined because there is no ceiling on sophistication or level of effort. The most common method used in Tier 3 is a performance evaluation using FEMA 356, *Prestandard and Commentary for the Seismic Rehabilitation of Buildings* (FEMA, 2000), based on simplified nonlinear analysis using pushover analysis (see Chapter 6). This method approximates the maximum lateral deformation that the building will suffer in a design event, considering the nonlinear behavior created by yielding and damage to components. The level of deformation of individual components is compared with standard deformations preset to performance levels of Collapse Prevention, Life Safety, and Immediate Occupancy, with the probable damage state of the building as a whole set at that level of the worst component. Efforts are being made to more realistically relate the damage states of all the components to a global damage state. This method can be used either to determine the probable damage state of the building for evaluation purposes, or to check the ability of a retrofit scheme to meet a target level.

With the advancement of computer capability and analysis software, nonlinear analysis techniques are constantly being improved. The ultimate goal, although not expected to be an everyday tool in the near future, is to simulate the movements of the full buildings during an entire earthquake, including the constantly changing properties of the structural components due to yielding and damage. The overall damage to various components is then accumulated and the global damage state thereby surmised.

### 8.4.3 Other Evaluation Issues

There are several other issues associated with seismic evaluation that should be recognized. Only three will be discussed here. First is the data required to perform competent evaluations at the various levels, as discussed above. Second, it is important to understand the performance expectation of the pass-fail line for various evaluation methodologies. Last, the reliability (or lack thereof) of the methods and of evaluation and/or performance prediction in general, should be recognized.

● Data Required for Seismic Evaluation

Obviously, for methods depending on the FEMA building type, the building type must be known. In fact, there are other similar classifications of building types also used to define building performance at the broadest level. Using data as discussed in paragraph 8.4.1, crude expectations of performance and therefore comparative evaluation can be completed. Most such systems, however, are refined by age, physical condition of the building, configuration, and other more detailed data, when available. Most "rapid" evaluation methods, based on building type and very basic building characteristics, do not require structural drawings. Responsible evaluators will insist on a site visit (in many cases to make sure the building is still there, if nothing else).

The more standardized evaluation methods discussed in paragraph 8.4.2 essentially require drawings. If detailed structural drawings are not available, simple evaluations of some model building types (wood buildings, tilt-ups, and sometimes URM) can be performed based on layout drawings or from data prepared from field visits. However, when reinforced concrete, reinforced masonry, or structural steel is a significant part of the structure, it is most often economically infeasible to reproduce "as-built" drawings. Practically in those cases, with rare exceptions, the building is deduced to be in nonconformance and, as a retrofit, a new seismic system is introduced to render the unknowns of the existing structure insignificant. Even in those cases, however, extensive field work is necessary to produce enough structural data to create a reasonable set of construction documents.

If original structural drawings are available that are confirmed to be reasonably accurate from spot checks in the field, most evaluation techniques can be employed. However, material properties are often not included on the drawings and must be deduced from the era of construction. Deterioration can also affect the material properties of several building types. Often the potential variability in the analysis due to different possible combinations of material properties requires in-situ testing of material properties. The techniques for this testing are well established, but cost and disruption to tenants are often an issue.

As explained in Chapter 3, "Site Evaluation and Selection", many areas of the country are mapped in detail for seismic parameters related to design, although such parameters continue to be investigated and updated. When warranted, site-specific studies can be performed to obtain timely and locally derived data. However, other seismic site hazards, such as liquefaction, landslide, and potential surface fault rupture, are less well mapped and may require a site-specific study, if there is reason to suspect their potential at a site.

Perhaps a less obvious important characteristic of a site is the detail of adjacent structures. Particularly in urban settings, adjacent buildings often have inadequate separation or are even connected to the building to be evaluated. Although legal issues abound when trying to deal with this issue, it is unrealistic to analyze and evaluate such a building as if it were freestanding. Formal evaluation techniques, such as ASCE 31, have addressed this issue, at least for buildings that are not connected, by highlighting the conditions known to potentially produce significant damage. First, if floors do not align between adjacent buildings and pounding is expected, the stiff floor from one building could cause a bearing wall or column in the adjacent building to collapse. Secondly, if buildings are of significantly different height, the interaction from pounding has been observed to cause damage. See Figure 8-8.

● Performance Objectives and Acceptability

Traditionally, evaluation techniques have been targeted at determining if a building is adequately life safe in an earthquake, similar to code goals for new buildings. However, as discussed in section 8.2.2, a standard different and less than that used for new buildings evolved, but was still termed life safety. Only with the development of performance-based engineering did evaluation methods aimed at other performance standards emerge. Even life safety has proven to be amorphous over the years and often has been defined by the evaluation technique du jour. Seismically, life safety is a difficult concept, due to the huge potential variation in ground motion and the many sources of damage that could cause injury

Figure 8-8: Typical adjacency issues in urban settings.

The small building in the center clearly cannot fail side to side, but this condition is not considered in mandatory retrofit ordinances that assume the adjacent buildings may be removed. In fact, however, the small building is in great danger from falling debris from its taller neighbors. In addition, the taller buildings are at risk from receiving serious damage to the corner columns from pounding against the shorter building.

or death. However, the term is well embedded in public policy and continues to persist in seismic codes and standards.

FEMA, in sponsoring the development of FEMA 273 (and later FEMA 356), wanted a more specific definition of a suitable goal for seismic safety, and thus the **Basic Safety Objective** (BSO) was defined. This performance objective consists of two requirements: the building would provide life safety for the standardized code event and, in addition, the building would not collapse in the **Maximum Considered Event** (MCE), a very rare event now defined by code. Since these FEMA documents are non-mandatory (unless locally adopted), the BSO has not become a widely accepted standard (the BSO also includes mandatory nonstructural minimum requirements, which also may delay its wide acceptance).

EXISTING BUILDINGS–EVALUATION AND RETROFIT

Chapter 6 contains a detailed discussion of performance-based engineering, which is gaining acceptance for evaluations at any level. But performance characterization in various forms has been used for some time, primarily to set policy. Such policies require descriptions of various performance levels, even if the technical ability to define or predict the various levels often lagged behind. Table 8-2 shows several such performance descriptions, many developed decades ago, which have been used to set policy—most concentrating on life safety. The table is set up to approximately equilibrate levels of performance across horizontal lines.

The first columns in Table 8-2 describe a system used by the University of California. GOOD, the best performance, is defined as the equivalent life safety as that provided by the code for new buildings—but without consideration of monetary damage. The next level below was set at the acceptable level for evaluation, while retrofits are required to meet the GOOD level.

The next column, labeled "DSA", is a Roman numeral system developed by the California Division of the State Architect for use with state-owned buildings. Each level has a description of damage and potential results of damage ("building not reoccupied for months") but no reference to engineering parameters. The state used an acceptance level of IV, but set the goal for retrofits to III.

The levels described in the next columns come from one of the early developments of performance based earthquake engineering, *Vision 2000*, developed by the Structural Engineers Association of California. It is a relatively comprehensive scale using five primary descriptions of damage, each with a "plus" and "minus", resulting in ten levels.

Finally, to indicate a perhaps more commonly recognized standard of performance, are the three occupancy tagging levels of Red, Yellow, and Green used for emergency evaluation immediately after damaging earthquakes.

● Reliability of Seismic Evaluations

The most significant characteristic in the design of buildings for earthquakes is the variability of ground motions. Not only do magnitudes and locations vary, but also the effects of fault rupture, wave path, and local

Table 8-2: A comparison of performance classifications

| COMPARATIVE CROSS REFERENCE OF VARIOUS SEISMIC PERFORMANCE RATINGS | | | | | | | |
| --- | --- | --- | --- | --- | --- | --- | --- |
| UC Rating | | | DSA* Rating | SEAOC Vision 2000 Rating | | | ATC-20 Post-Earthquake |
| Designation | Structural Damage Level | Hazard to Life | | Numerical Rating | Performance Expectation | Anticipated Damage | Assessment Designation |
| GOOD | Some Damage | Not Significant | I | 10 | Fully Operation | Negligible | "Green Tag" Inspected No Restriction on Use or Occupancy |
| | | | II | 9 | | | |
| | | | II | 8 | Operational | Light | |
| | | | III | 7 | | | |
| | | | III | 6 | Life Safety | Moderate | |
| | | | IV | 5 | | | "Yellow Tag" Limited Entry Off Limits to Unauthorized Personnel |
| FAIR | Damage | Low | IV | 4 | Near Collapse | Severe | |
| POOR | Significant Damage | Appreciable | V | 3 | | | |
| | | | VI | 2 | Partial Collapse | | |
| VERY POOR | Extensive Damage | High | VI | 2 | Partial Collapse (Large Assembly Area Only) | Complete | "Red Tag" Unsafe Do Not Enter or Occupy |
| | | | VII | 1 | Total Collapse | | |

site soils create a literally infinite set of possible time histories of motion. Studies have shown that time histories within a common family of parameters used for design (response spectrum) can produce significantly different responses. This variation normally dominates over the scatter of results from analysis or evaluation techniques.

However, codes for new buildings can require many limitations of material, lateral system, configuration, and height that will reasonably assure acceptable performance, particularly the prevention of collapse. These same limitations can seldom be applied to existing buildings, so the variation of actual performance is expected to be much larger. In addition, the cost of retrofit is often high, and attempts have been made to avoid unnecessary conservatism in evaluation methodologies. It is probable, therefore, that a significant number of buildings may fail to perform as evaluated, perhaps in the range of 10% or more. No comprehensive study has been made to determine this reliability, but ongoing programs to further develop performance-based seismic engineering are expected to estimate the variability of evaluation results and refine the methods accordingly.

## Box: Describing Seismic Performance

Seismic performance is specified by selecting a maximum tolerable damage level for a given earthquake-shaking intensity. The shaking intensity can be specified **probabilistically**, derived by considering all future potential shaking at the site regardless of the causative fault, or **deterministically**, giving the expected shaking at the site for a given-sized earthquake on a given fault. The damage level can be described using one of several existing scales, including the DSA Risk Levels or performance levels developed in the long-running FEMA program to mitigate seismic risks from existing buildings.

## Describing Shaking Intensity

For some time, the earthquake shaking used by the building code for new buildings has been described probabilistically, as shaking with a 10% chance of being exceeded in a 50-year time period (50 years being judged as the average life of buildings). This can also be specified, similar to methods used with storms or floods, as the shaking with a return period of 475 years. (Actually, for ease of use, the return period is often rounded to 500 years, and since actual earthquake events are more understandable than probabilistic shaking, the most common term, although slightly inaccurate, is "the 500-year event".)

Nationally applicable building codes were therefore based on the level of shaking intensity expected at any site once every 500 years (on average). However, engineers in several areas of the country, most notably Salt Lake City, Utah; Charleston, South Carolina; and Memphis, Tennessee, felt that this standard was not providing sufficient safety in their regions because very rare, exceptionally large earthquakes could occur in those areas, producing shaking intensities several times that of the 500 year event. Should such a rare earthquake occur, the building code design would not provide the same level of protection provided in areas of high seismicity, particularly California, because rare, exceptionally large shaking in California is estimated to be only marginally larger (about 1.5 times) than the 500-year shaking. It was therefore decided to determine the national mapping parameters on a much longer return period—one that would capture the rare events in the regions at issue, and a 2,500 year event was chosen (known as the Maximum Considered Event—MCE). Finally, it was judged unnecessary, and in fact undesirable, to significantly change seismic design practices in California, so the MCE was multiplied by 2/3 to make California design shaking levels about the same as before.

continued next page

(If the new shaking level - about 1.5 times the old - were multiplied by 2/3, the final design parameter would not change.) However, in a region where the MCE is 3 times the previously used 500-year event, the new parameter of 2/3 MCE would result in a shaking level twice that previously used—providing the sought-after additional level of safety in those regions. Currently, national standards such as ASCE 31 define the level of shaking to be considered for evaluation of existing buildings to be 2/3 MCE, which, as previously explained, is about the same as the 500-year event for much of California.

## Describing Damage Levels

Although several descriptions of performance damage levels are currently in use (see Table 8-2), descriptions of FEMA performance levels summarized from FEMA 356 (FEMA, 2000), which covers the full range of performance, are given below:

- **Operational**: Buildings meeting this performance level are expected to sustain minimal or no damage to their structural and nonstructural components. The building will be suitable for its normal occupancy and use, although possibly in a slightly impaired mode, with power, water, and other required utilities provided from emergency sources. The risk to life safety is extremely low.

- **Immediate Occupancy**: Buildings meeting this performance level are expected to sustain minimal or no damage to their structural elements and only minor damage to their nonstructural components. Although immediate re-occupancy of the building will be possible, it may be necessary to perform some cleanup and repair and await the restoration of utility service to function in a normal mode. The risk to life safety is very low.

- **Life Safety**: Buildings meeting this performance level may experience extensive damage to structural and nonstructural components. Structural repair may be required before re-occupancy, and the combination of structural and nonstructural repairs may be deemed economically impractical. The risk to life safety is low.

- **Collapse Prevention**: Buildings meeting this performance level will not suffer complete or partial collapse nor drop massive portions of their structural or cladding on to the adjacent property. Internal damage may be severe, including local structural and nonstructural damage that poses risk to life safety. However, because the building itself does not collapse, gross loss of life is avoided. Many buildings in this damage state will be a complete economic loss.

## 8.5 SEISMIC REHABILITATION OF EXISTING BUILDINGS

There are many reasons why buildings might be seismically retrofitted, including renovations that trigger a mandatory upgrade, a building subjected to a retroactive ordinance, or the owner simply wanting (or needing) improved performance. The reason for the upgrade may influence the technique and thoroughness of the work, because owners faced with mandatory upgrade may seek out the least expensive, but approvable, solution, whereas an owner needing better performance will more likely be willing to invest more for a solution that addresses their particular concerns. There are many other factors that shape a retrofit solution, such as the type of deficiency present, if the building is occupied, and the future use and aesthetic character of the building.

The continuing improvement of analysis techniques and the emergence of performance-based design are also having a large effect on retrofit schemes, by enabling engineers to refine their designs to address the specific deficiencies at the desired performance level. In many cases, however, the retrofits are becoming controlled by the brittleness of existing components that must be protected from excess deformation with systems that may be stronger and/or stiffer than those used for new buildings. Some older retrofits, done to prescriptive standards or using now-outdated strengthening elements borrowed from new building designs, may themselves be deficient, depending on the desired performance. Seismic retrofit analysis, techniques, and components, similar to new building technology, are not static, and applications should be regularly reviewed for continued effectiveness.

### 8.5.1 Categories of Rehabilitation Activity

In most cases, the primary focus for determining a viable retrofit scheme is on vertically oriented components (e.g. column, walls, braces, etc.) because of their significance in providing either lateral stability or gravity-load resistance. Deficiencies in vertical elements are caused by excessive inter-story deformations that either create unacceptable force or deformation demands. However, depending on the building type, the walls and columns may be adequate for seismic and gravity loads, but the building is inadequately tied together, still forming a threat for partial or complete collapse in an earthquake. It is imperative to have a thorough understanding of the expected seismic response of the existing building, and all of its deficiencies to design an efficient retrofit scheme. There are three basic categories of measures taken to retrofit a building:

1) Modification of global behavior, usually decreasing deformations (drifts);

2) Modification of local behavior, usually increasing deformation capacity;

3) Connectivity, consisting of assuring that individual elements do not become detached and fall, assuring a complete load path, and assuring that the force distributions assumed by the designer can occur.

The types of retrofit measures often balance one another, in that employing more of one will mean less of another is needed. It is obvious that providing added global stiffness will require less deformation capacity for local elements (e.g. individual columns), but it is often less obvious that careful placement of new lateral elements may minimize a connectivity issue such as a diaphragm deficiency. Important connectivity issues such as wall-to-floor ties, however, are often independent and must be adequately supplied.

● Modification of Global Behavior

Modification to global behavior normally focuses on deformation, although when designing to prescriptive standards, this may take the form of adding strength. Overall seismic deformation demand can be reduced by adding stiffness in the form of shear walls or braced frames. Addition of moment frames is normally ineffective in adding stiffness. New elements may be added or created from a composite of new and old components. Examples of such composites include filling in openings of walls and using existing columns for chord members for new shear walls or braced frames.

Particular ground motions have a very specific deformation demand on structures with various periods, as discussed in Chapter 4. Given an equal period of vibration, this deformation will occur, whether distributed over the height of the building or concentrated at one floor. If one or more inter-story drifts are unacceptable, it may be possible to redistribute stiffness vertically to obtain a more even distribution of drift. A soft or weak story is an extreme example of such a problem. Such stories are usually eliminated by adding strength and stiffness in such a way as to more closely balance the stiffness of each level, and thus evenly spread the deformation demand over the height of the structure.

Seismic isolation is the supreme example of the concept of redistribution of deformation. Essentially all deformation is shifted to bearings, placed at the isolation level, that are specifically designed for such response. The bearings limit the response of the superstructure, which can be designed to remain essentially undamaged for this maximum load. The feasibility of providing isolation bearings that limit superstructure accelerations to low levels not only facilitates design of superstructures to remain nearly elastic, but also provides a controlled environment for design of nonstructural systems and contents.

Global deformations can also be controlled by the addition of passive energy dissipation devices, or **dampers**, to the structure. Although effective at controlling deformations, large local forces may be generated at the dampers that must be transferred from the device to structure and foundation, and the disruptive effect of these elements on the interior of the building is no different than a rigid brace.

## ● Modification of Local Behavior

Rather than providing retrofit measures that affect the entire structure, deficiencies also can be eliminated at the local component level. This can be done by enhancing the existing shear or moment strength of an element, or simply by altering the element in a way that allows additional deformation without compromising vertical-load carrying capacity.

Given that in most cases, that certain components of the structure will yield (i.e., become inelastic), some yielding sequences are almost always benign: beams yielding before columns, bracing members yielding before connections, and bending yielding before shear failure in columns and walls. These relationships can be determined by analysis and controlled by local retrofit in a variety of ways. Columns in frames and connections in braces can be strengthened, and the shear capacity of columns and walls can be enhanced to be stronger than the shear that can be delivered.

Concrete columns can be wrapped with steel, concrete, or other materials to provide confinement and shear strength. Concrete and masonry walls can be layered with reinforced concrete, plate steel, and other materials. Composites of glass or carbon fibers and epoxy are becoming popular to enhance shear strength and confinement in columns, and to provide strengthening to walls.

Another method to protect against the collapse risk posed by excess drift is to provide a supplementary gravity support system for elements that might be unreliable at expected high-deformation levels. For example, supplementary support for concentrated wall-supported loads is a requirement in California standards for retrofit of unreinforced masonry buildings. In several cases, supplementary support has also been used in concrete buildings.

Lastly, deformation capacity can be enhanced locally by uncoupling brittle elements from the deforming structure, or by removing them completely. Examples of this procedure include placement of vertical saw cuts in unreinforced masonry walls to change their behavior from shear failure to a more acceptable rocking mode, and to create slots between spandrel beams and columns to prevent the column from acting as a "short column" prone to shear failure.

● Connectivity

Connectivity deficiencies are within the load path: wall out-of-plane connection to diaphragms; connection of diaphragm to vertical lateral force-resisting elements; connection of vertical elements to foundation; connection of foundation to soil. A complete load path of some minimum strength is always required, so connectivity deficiencies are usually a matter of degree. A building with a complete but relatively weak or brittle load path might be a candidate for retrofit by seismic isolation to simply keep the load below the brittle range.

The only location in the connectivity load path at which yielding is generally allowed is the foundation/structure interface. Allowing no movement at this location is expensive and often counterproductive, as fixed foundations transfer larger seismic demands to the superstructure. Most recently developed retrofit guidelines are attempting to provide simplified guidance to the designer on how to deal with this difficult issue and minimize foundation costs.

### 8.5.2 Conceptual Design of a Retrofit Scheme for an Individual Building

There are many specific methods of intervention available to retrofit designers, as previously discussed. The selection of the specific type of element or system is dependent on local cost, availability, and suitability for the structure in question. Any system used to resist lateral load in new buildings can also be used for retrofit. It is thus an extensive task to develop guidelines for such selection. In addition, as in the design of a new building, there is usually a choice of where to locate elements, although it is generally more restrictive in existing buildings. However, in the end, there are nonseismic issues associated with each building or project that most often control the specific scheme to be used.

The solution chosen for retrofit is almost always dictated by building user-oriented issues rather than by merely satisfying technical demands. There are five basic issues that are always of concern to building owners or users: **seismic performance, construction cost, disruption to the building users during construction** (often translating to a cost), **long-term affect on building space planning**, and **aesthetics**, including consideration of historic preservation.

All of these characteristics are always considered, but an importance will eventually be put on each of them, either consciously or subconsciously, and these weighting factors invariably will determine the scheme chosen.

○ Seismic performance

Prior to the emphasis on performance-based design, perceived qualitative differences between the probable performance of difference schemes were used to assist in choosing a scheme. Now, specific performance objectives are often set prior to beginning development of schemes. Objectives that require a very limited amount of damage or "continued occupancy" will severely limit the retrofit methods that can be used and may control the other four issues.

○ Construction cost

Construction cost is always important and is balanced against one or more other considerations deemed significant. However, sometimes other economic considerations, such as the cost of disruption to building users, or the value of contents to be seismically protected, can be orders of magnitude larger than construction costs, thus lessening its importance.

○ Disruption to the building users during construction

Retrofits are often done at the time of major building remodels, and this issue is minimized. However, in cases where the building is partially or completely occupied, this parameter commonly becomes dominant and controls the design.

○ Long-term effect on building space planning

This characteristic is often judged less important that the other four and is therefore usually sacrificed to satisfy other goals. In many cases, the planning flexibility is only subtly changed. However, it can be significant in building occupancies that need open spaces, such as retail spaces and parking garages.

○ Aesthetics

In historic buildings, considerations of preservation of historic fabric usually control the design. In many cases, even performance objectives are controlled by guidelines imposed by preservation. In non historic buildings, aesthetics is commonly stated as a criterion, but in the end is often sacrificed, particularly in favor of minimizing cost and disruption to tenants.

These parameters can merely be recognized as significant influences on the retrofit scheme or can be used formally to compare schemes. For example, a comparison matrix can be developed by scoring alternative schemes in each category and then applying a weighting factor deduced from the owner's needs to each category.

Figure 8-9 describes the evolution of a retrofit scheme based on several changes in the owner's weighting of these five characteristics.

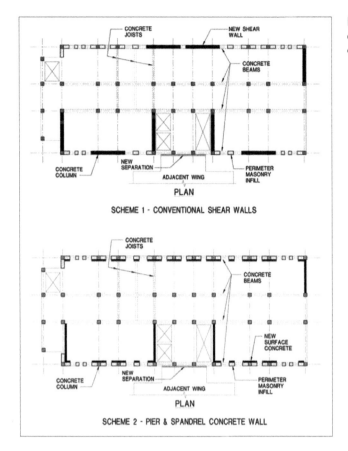

Figure 8-9a: Example of effect of non technical issues on retrofit schemes.

This is a seven-story concrete building built in the early 1920s. It consists of two wings in a T shape. The plans show the second of the two wings poorly connected at the location indicated. The building is a concrete frame with brick infill exterior walls, and lateral forces were not apparently considered in the original design. Although not officially judged historic, the exterior was articulated and considered pleasing and a good representative of its construction era. As can be seen in the plan, the building has no lateral strength in the transverse direction other than the poor connection to the second wing, and was evaluated to present a high risk to occupants.

It was judged early that the vertical load carrying elements had little drift capacity, and that stiffening with shear walls was the only feasible solution. The first two schemes shown are straightforward applications of shear walls. The first concentrates the work to minimize disruption, but closes windows and creates large overturning moments at the base. The second distributes the longitudinal elements and preserves windows by using a pier-spandrel shear wall. Both schemes separated the wings into two buildings, to allow future demolition of either to facilitate phasing for a new replacement building sometime in the future. The cost and disruption was judged high, and the work would have to be phased upwards, evacuating three floors at a time to avoid the noise and disruption.

Figure 8-9b: Example of effect of non technical issues on retrofit schemes.

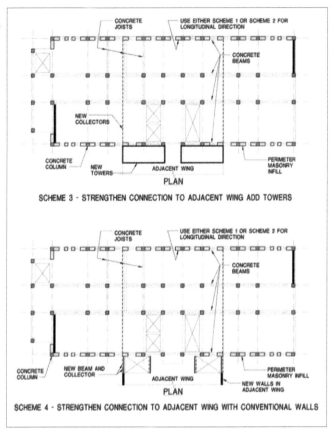

SCHEME 3 - STRENGTHEN CONNECTION TO ADJACENT WING ADD TOWERS

SCHEME 4 - STRENGTHEN CONNECTION TO ADJACENT WING WITH CONVENTIONAL WALLS

Schemes 1 and 2 required a complete lateral system in both buildings because of the separation installed at the wing intersection, which also caused difficult exiting issues. Schemes 3 and 4 were therefore developed, providing a strong inter-tie between wings and taking advantage of several new lateral elements to provide support to both wings. Scheme 3 featured new concrete towers as shown. Although the outside location was considered advantageous from a disruption standpoint, the towers closed windows and caused disruption to mechanical services. Scheme 4 was similar to 3, but eliminated the towers. Schemes 3 and 4 had less construction cost than 1 and 2, but disruption, in terms of phasing, caused essentially the same total downtime.

EXISTING BUILDINGS–EVALUATION AND RETROFIT

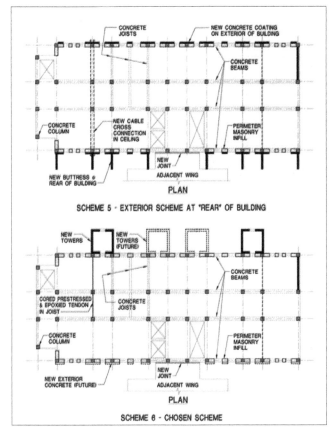

SCHEME 5 - EXTERIOR SCHEME AT "REAR" OF BUILDING

SCHEME 6 - CHOSEN SCHEME

Figure 8-9c: Example of effect of non technical issues on retrofit schemes.

When the owner completed a study of the availability and cost of surge space in the area to facilitate the phasing required for Schemes 1–4, it was discovered that the cost of moving and rental space was larger by far than the construction costs thus far budgeted.

Occupant disruption thus became the primary control parameter for development of retrofit schemes, and aesthetics, measured by preservation of the exterior appearance, was significantly reduced as a consideration. Scheme 5 was developed with buttresses on the off-street side of the building and longitudinal walls applied from the outside. Collectors to the buttresses were to be post-tensioned cables installed through conduit placed at night in the ceiling spaces. Access to the rear of the building was difficult, so aesthetic considerations were further relaxed to allow buttresses—that became towers—on the front side. Collectors to the towers were post-tensioned rods installed in cores drilled approximately 16 feet (5 m) into the building in the center of 12-inch by 16-inch (30x40cm) joists. Both Schemes 5 and 6 installed a seismic separation at the intersection of wings to facilitate partial replacement. At the request of the contractor, another scheme, 6a, was developed that replaced the concrete shear walls with steel-braced frames, but it proved no more economical. Scheme 6 was selected for construction, although – since replacement at that time appeared to be in the long range planning stages –  the owner chose to construct only part of Scheme 6, aimed at eliminating the obvious collapse mechanism in the transverse direction. As shown, only two towers were constructed, and the only longitudinal strengthening provided was the weak-way strength of the towers.

Figure 8-10: Intense construction activity and disruption from interior shotcrete.

SOURCE: RUTHERFORD & CHEKENE, CONSULTING ENGINEERS

Figure 8-11: Retrofit activities inside buildings are most often not surgical or delicate. Here work on a new foundation for a shear wall is prepared for casting.

SOURCE: RUTHERFORD & CHEKENE, CONSULTING ENGINEERS

### 8.5.3 Other Rehabilitation Issues

● Inadequate recognition of disruption to occupants

It is unfortunately common for the extent of interior construction and disruption to be underestimated. In many cases, occupants who were originally scheduled to remain in place are temporarily moved—at a significant increase in cost of the project—or the work is required to be done in off-hours, also a premium cost. Figures 8-10 and 8-11 indicate the level of construction intensity often required in retrofit.

Similarly, "exterior solutions," where strengthening elements are placed on the outside of the building are often more disruptive and noisier than

Figure 8-12: External towers were added to strengthen this building from the outside. The new tower is the element at center-left of the figure. The retrofit scheme for this building is discussed in detail in Figure 8-9.

anticipated and often require collector members to be placed on each floor within the building. Figure 8-12 shows the result of an exterior retrofit of adding towers on the outside of a building that, in fact, did not cause a single lost day of occupancy. High-strength steel rods were epoxied into horizontal cores, drilled twenty feet into the existing concrete beams to form the needed collectors.

● Collateral required work

As previously mentioned, retrofit work is often performed in conjunction with other remodeling or upgrading activities in a building. Such work normally triggers other mandatory improvements to the building, such as ADA compliance or life safety updating—all of which add cost to the project. However, even when seismic retrofit is undertaken by itself, the costs of ADA compliance, removal of disturbed hazardous material, and possibly life safety upgrades must be considered.

### 8.5.4 Examples

It is impossible to include examples that show the full range of structural elements and configurations used in seismic retrofit. There are definitely patterns, usually driven by economics or avoidance of disruption to occupants, but depending on the particular mix of owner requirements, as discussed in Section 8.5.2, thoughtful architects and engineers will always come up with a new solution.

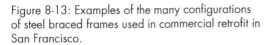

Figure 8-13: Examples of the many configurations of steel braced frames used in commercial retrofit in San Francisco.

The retrofits were probably required by the URM Ordinance or triggered by upgrading or remodeling. Tall narrow brace configurations, as shown in the upper left and lower right, are less efficient that more flat brace orientations.

Figure 8-14: Examples of steel moment frames in similar commercial retrofits. Right: Note the large white column and double beam arrangement. Left: The moment frame is placed against the wall of the recess. The first floor columns are gray and the balance is pink. The frame can be seen on both the first and second floors.

Figure 8-15: Renovation and retrofit of a concrete warehouse structure by removing the exterior wall, inserting steel braces and window wall, and adding several floors.

Figure 8-16: Steel braced frames on exterior of building to avoid construction on the inside of the building.

Figure 8-17: A modern exterior buttress used for seismic strengthening.

SOURCE: TAKANE ESHIMA, MURAKAMI/NELSON

Figure 8-18: Infill of certain panels of an exterior wall for strengthening. Originally on the alley wall, only the central stair tower was solid wall, and elsewhere the upper window panel pattern was typical to the street.

Figure 8-19: Examples of addition of a new wall on the exterior of the building.

Left: The end wall has a new layer of concrete that wraps around the side for a short distance. This solution is unusual in an urban setting because of property lines. Right: a large academic building with more extensive C-shaped end elements. This solution facilitated construction while the building was occupied, although there was considerable noise and disruption experienced by the occupants.

## Box 3: CASE STUDY: SEISMIC RETROFIT BREMERTON NAVAL HOSPITAL WASHINGTON

The U.S. Navy recognized in the late 1990s that the Bremerton Naval Hospital in Bremerton, Washington, was important not only for the 60,000 military families in the area, but also that it might be called upon to serve more than 250,000 people in the immediate area in the event of a major earthquake. Accordingly, a detailed seismic evaluation of the hospital using performance-based design engineering standards (FEMA 310, *Handbook for the Seismic Evaluation of Buildings*, A Prestandard and FEMA 356, *Prestandard and Commentary for the Seismic Rehabilitation of Buildings*) was performed to gain a better understanding of the potential seismic deficiencies.

The building's lateral-resisting system, constructed in the "pre-Northridge" 1960s, is comprised of steel moment-resistant frames at all beam-column connections. Although highly redundant, it is too flexible, resulting in excessive drift. The cladding panel connections were not designed to accommodate the expected drifts from a design-level earthquake, and presented a potential falling hazard. Additionally, there was incompatibility between the flexible structure and the rigid concrete stair tower.

## BREMERTON NAVAL HOSPITAL continued

This detailed evaluation was completed in late 2001. In February 2002, the magnitude 6.8 Nisqually earthquake shook the Puget Sound region. Shaking at the hospital was modest, because the earthquake epicenter was located approximately 30 miles away. A seismograph at the hospital recorded a horizontal peak acceleration of 0.11g at the basement level and a peak roof acceleration of 0.47g. Calculated peak roof displacements from this modest earthquake were over 6 inches (a floor-to-floor drift ratio of 0.5%).

Because a traditional seismic retrofit that strengthened and stiffened the moment frames would have been costly and disruptive, alternate retrofit design methods were evaluated.

The use of supplemental passive damping devices proved to be the best approach to improve the seismic performance of the structure by reducing drift, while minimizing disruption during construction. Seismic forces, displacements, and floor accelerations would be substantially reduced by dissipation of the earthquake's energy through heat created in the damping devices. A total of 88 seismic dampers were installed at 44 select locations in the building.

Target performance levels were "Immediate Occupancy" for the 10%/50-year Design Basis Earthquake (DBE) and "Collapse Prevention" for the 2%/50-year Maximum Considered Earthquake (MCE).

CREDITS:CASE STUDY BASED ON THE ARTICLE STRONG MEDICINE, AUTHORS DOUGLAS WILSON, PE: RUSSELL KENT, PE; STEPHAN STANEK, PE AND DAVID SWANSON, PE,SE; IN MODERN STEEL CONSTRUCTION, AMERICAN INSTITUTE OF STEEL CONSTRUCTION, CHICAGO, IL, FEBRUARY 2005.

Photographs of retrofit buildings, although often interesting, seldom can tell the full story of the development of the scheme, and if the majority of retrofit elements are inside or hidden, tell almost nothing. Some photos are shown here, but are not intended to demonstrate the full range of buildings that have successfully undergone seismic retrofit or the full range of solutions to individual problems. In addition, due to limited space, only one or two points are made with each photo, rather than a full case study.

## 8.6 SPECIAL ISSUES WITH HISTORIC BUILDINGS

Seismic evaluation and retrofit of historic buildings generate complex public policy issues for which few general rules can be identified. Restoration, or renovations of large and important historic buildings usually have considerable public and jurisdictional oversight, in addition to employing an experienced design team that includes a special historic preservation consultant. The control and oversight for less important buildings that have historic status at some level, or that may qualify for such status, are highly variable. Designers are cautioned to locally investigate approval procedures for alterations on such buildings as well as seismic requirement, for them.

### 8.6.1 Special Seismic Considerations

It has been recognized in most areas of high seismicity that local public policy concerning seismic retrofit triggers must include special considerations for historic buildings. As discussed in Section 8.2, initial seismic safety criteria for existing buildings were focused on requirements for new buildings, which were marginally appropriate for most older buildings, but completely inappropriate for historic buildings. Special allowances were therefore created for archaic materials that were not allowed in new buildings, and the overall seismic upgrade level was lowered to reduce work that could compromise historic integrity and fabric. These kinds of technical criteria issues have been somewhat mitigated by the completion of FEMA 356 and the emergence of performance-based earthquake engineering, because consideration of archaic materials and fine-tuning of performance levels are now part of the normal lexicon.

### 8.6.2 Common Issues of Tradeoffs

Many buildings in this country that qualify for historic status are not exceptionally old and can be made commercially viable. The changes that are needed for successful adaptive reuse will often conflict with strict preservation guidelines, and compromises are needed in both directions to achieve a successful project that, in the end, could save the building from continuing decay and make it more accessible to the public. These tradeoffs occur in many areas of design, but seismic upgrading work often requires interventions that are not needed for any other reason. These interventions often fall under historic preservation guidelines that call for clear differentiation of new structural components, or that discourage recreation of historic components that are removed. As previously indicated, there are no rules for these conditions, and the most appropriate solution for each case must be determined individually.

Another common conflict is between current preservation of historic fabric and future preservation of the building due to the chosen seismic performance level. Typically, a better target performance in the future, possibly preventing unrecoverable damage, requires more seismic renovation work now. Most historic preservation codes allow lower expected seismic performance to reduce construction work and minimize damage. Like many seismic policies, there have not been enough earthquakes with seismically damaged historic buildings to test this general philosophy. In an ever-growing number of cases of important buildings, this dilemma has been addressed using seismic isolation—which by reducing loading to the superstructure, reduces required construction work and also reduces expected damage in future earthquakes. Typically, however, installing isolation into an existing building is expensive and may require a significant public subsidy to make viable. Several high-profile city hall buildings such as San Francisco, Oakland, Berkeley (after the 1989 Loma Prieta earthquake), and Los Angeles (after the 1994 Northridge earthquake) have been isolated, with FEMA assistance as part of post earthquake damage repairs.

### 8.6.3 Examples of Historical Buildings

The following illustrations show samples of seismic retrofit of historic buildings with brief descriptive notes. Complete discussion of the preservation issues and rehabilitation techniques of each case would be extensive and cannot be included here.

Figure 8-20: Several examples of historical buildings seismically retrofit and protected by seismic isolation.

Upper left: Oakland City Hall (Engineer: Forell/Elsesser). Upper right: San Francisco City Hall (Engineer: Forell/Elsesser). Bottom: Hearst Mining Building at University of California, Berkeley (Engineer: Rutherford & Chekene). Installation of an isolation system under an existing building is complex and often expensive, but the system minimizes the need to disrupt historic fabric in the superstructure with shear walls or braces, and is designed to protect the superstructure from significant damage in a major event.

Figure 8-21: Mills Hall, Mills College, California. Structure is wood studs and sheathing. Interior shear walls were created by installing plywood on selected interior surfaces.

SOURCE: RUTHERFORD & CHEKENE, CONSULTING ENGINEERS

Figure 8-23: University Hall, University of California, Berkeley. The structure is unreinforced brick masonry bearing wall with wood floors and roof—a classic unreinforced masonry building (URM). Lateral resistance was added with interior concrete shear walls, improvements to the floor and roof diaphragms, and substantial ties from the exterior walls to the diaphragms.

SOURCE: RUTHERFORD & CHEKENE, CONSULTING ENGINEERS

Figure 8-22: St. Dominic's Church in San Francisco was seismically retrofitted using exterior buttresses.

SOURCE: RUTHERFORD & CHEKENE, CONSULTING ENGINEERS

Figure 8-24: Original Quad arcades, Stanford University. The Quad, approximately 850 feet by 950 feet in plan, is surrounded by a covered arcade. Modern seismic retrofits have taken place over a 40-year period, with evolving techniques. Final sections were completed by removal of the interior wythe of sandstone, installation of reinforced concrete core, and reinstallation of the interior blocks, which were reduced in thickness. Due to environmental decay of many of the sandstone columns, additional seismic resistance was obtained in many locations by installation of precast concrete replicas that formed a part of a continuous vertical reinforced concrete member.

SOURCE: RUTHERFORD & CHEKENE, CONSULTING ENGINEERS

Figure 8-25: Stanford University Quad Corner Buildings. The four very similar corner buildings were also seismically strengthened over a 40-year period. Substantially different techniques evolved due to growing recognition of the historic value of the entire Quad. The first corner was "gutted" in 1962 and an entire new structure built inside with two a,dded floors. The second corner was done in 1977 and was also gutted, but the original floor levels were maintained and interior finishes were similar to the original. The last two corners, repaired and retrofitted following the 1989 Loma Prieta earthquake, were strengthened with interior concrete shear walls, while the bulk of the interior construction was maintained, including the wood floors and heavy timber roofs.

SOURCE: RUTHERFORD & CHEKENE, CONSULTING ENGINEERS

Figure 8-26: The museum at San Gabriel Mission, southern California, a historic adobe structure damaged by earthquakes in 1987 and 1994.It was repaired and strengthened, including installation of a bond beam of steel in the attic, anchor bolts into the wall, and stitching of cracks back together on the interior walls.

ENGINEER: MEL GREENE AND ASSOC.

## Box 4: CASE STUDY: THE SEISMIC RETROFIT OF THE SALT LAKE CITY AND COUNTY BUILDING

The Salt Lake City and County Building was completed in 1894. It is now the seat of government for Salt Lake City, Utah. The historic landmark also housed offices for Salt Lake County until the 1980s but still retains its original name. The building is an unreinforced masonry bearing-wall structure with a 250-foot-high central clock tower (Figure A).

Figure A: The Salt Lake City and County Building.

By the 1970s, it had become obvious that substantial repairs were needed if the building was to be of further service, and limited restoration was done between 1973 and 1979. However, the building remained a potential source of injuries and costly lawsuits, and in the 1980s much public controversy began regarding whether the building should be demolished or saved and restored. After many architectural end engineering studies, in 1986 the city council approved financing for the restoration of the building.

After extensive materials testing and structural analysis, the decision was made to use a base-isolation scheme, consisting of over 400 isolators that would be installed on top of the original strip footings, with a new concrete structural system built above the bearings to distribute loads to the isolators (Figure B).

Calculations showed a dramatic reduction in forced levels in the superstructure, so that shotcreting of existing walls would not be necessary, and consequently the historic interior finishes could be saved. Existing floor diaphragms would require only minimum strengthening around their perimeters. However, due to the significance of the towers as a seismic hazard, it was decided to use the results of conventional non-isolated analysis for sizing the steel space frame members used for the tower strengthening (Figure C ).

Predicted damage from future earthquakes would also be substantially reduced, providing greatly increased safety to the building occupants. It would be necessary, however, to entirely

Figure B: Plan of isolator locations in basement.

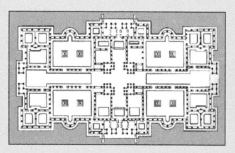

SOURCE: *SEISMIC ISOLATION RETROFITTING*, BY JAMES BAILEY AND EDMUND ALLEN, REPRINTED FROM THE VOLUME XX, 1988 ISSUE OF THE APT BULLETIN, THE JOURNAL OF THE ASSOCIATION FOR PRESERVATION TECHNOLOGY INTERNATIONAL

remove the first floor in order to provide space for the foundation work. In effect, the solution shifted the focus of the structural work from the shear walls above to the foundation, reducing much of the seismic retrofit work to a massive underpinning project.

Typical bearings used were approximately 17 inches square by 15 inches high and consisted of alternating layers of steel and rubber bonded together with a lead core (Figure D).

Figure C: Steel space frame inserted within tower

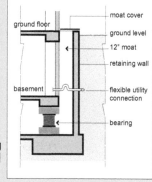

Figure D: Typical bearing layout with retaining wall and moat.

For the bearings to work properly, it was necessary to isolate the building from the ground horizontally. To do this, a retaining wall was constructed around the building perimeter with a 12-inch seismic gap (or moat), so that the building was free to move relative to the surrounding ground.

The retrofit of the Salt Lake City and County Building was completed in 1989 and was the world's first application of seismic base isolation for a historical structure.

CREDITS: PROJECT ARCHITECT: THE EHRENKRANTZ GROUP, NEW YORK. ASSOCIATE ARCHITECT: BURTCH BEALL JR. FAIA, SALT LAKE CITY, UT. STRUCTURAL ENGINEER: E.W. ALLEN AND ASSOCIATES, SALT LAKE CITY, UT. BASE ISOLATION CONSULTANT: FORELL/ELSESSER, SAN FRANCISCO, CA.

## 8.7 CONCLUSION

Table 8-3 summarizes common seismic deficiencies stemming from various site and configuration characteristics as well as those that might be expected in each FEMA model building type. See Section 8.2 for a discussion of "seismic deficiency" as used in this chapter and this table. Also included in Table 8-3 are retrofit measures that are often used for each situation.

## 8.8 REFERENCES

### 8.8.1 References from Text

American Society of Civil Engineers (ASCE), *Seismic Evaluation of Existing Buildings*, ASCE 31-03, 2003.

Applied Technology Council (ATC), *Earthquake Damage Evaluation Data for California*, ATC 13, 1985.

California Seismic Safety Commission (CSSC), *Status of the Unreinforced Masonry Builidng Law*, 2003 Report to the Legislature, SSC 2003-03, 2003.

Federal Emergency Management Agency (FEMA), *Prestandard and Commentary for the Seismic Rehabilitation of Buildings*, FEMA 356, 2000.

Holmes, William T., Background and History of the California Hospital Seismic Safety Program, *Proceedings, Seventh National Conference on Earthquake Engineering*, Boston, 2002, Earthquake Engineering Research Institute.

Hoover, Cynthia A., *Seismic Retrofit Policies: An Evaluation of Local Practices in Zone 4 and Their Application to Zone 3*, 1992, Earthquake Engineering Research Institute.

National Institute of Standards and Technology (NIST), *Standards of Seismic Safety for Existing Federally Owned or Leased Buildings*, ICSSC RP 6, 2002.

U. S. Department of Interior (Interior a), *The Secretary of the Interior's Standards for Rehabilitation, Department of Interior Regulations, 36 CFR 67*, U. S Department of the Interior, National Park Service, Preservation Assistance Division, Washington, D.C.

U. S. Department of Interior (Interior b), *Guidelines for Rehabilitating Historic Buildings*, U. S Department of the Interior, National Park Service, Preservation Assistance Division, 1977, Washington, D.C.

Table 8-3: Common deficiencies by category of buildings

| Category | Physical Characteristic | Performance Characteristics | Common Retrofit Techniques |
|---|---|---|---|
| Site | Liquefaction | • Small settlements from thin layers of liquefied material.<br>• Large settlements and/or loss of foundation support from thick layers.<br>• Horizontal flow possible if massive amounts of material liquefy, even with slight slopes. | • Stabilize soil with cement injection or by draining water to eliminate saturated state.<br>• Place building on deep foundation that can withstand layer of liquefied material. |
| | Potential fault rupture | • If fault rupture is very near but not through building, only unique effect may be broken utilities or loss of access.<br>• If fault rupture passes through building, severe damage to building is likely. | • Avoid condition if possible.<br>• Retrofit possible in some cases with massive foundation that will force fault slippage around or under building without causing collapse of superstructure. |
| | Adjacent buildings | • If contact is expected, short adjacent buildings can cause a soft-story effect on the levels immediately above the short building.<br>• If floors do not align, load-bearing columns or walls can be damaged by pounding, potentially causing collapse.<br>• If walls share a structural wall ("common" wall), interaction may be extreme, and individual analysis is required.<br>• Taller buildings, particularly URMs or buildings with URM exterior walls or parapets, may drop debris on shorter buildings, that potentially will pass through the roof. | • These conditions are difficult to mitigate without cooperation from both property owners.<br>• Potential contact areas can be strengthened, but this may cause additional damage to neighboring building.<br>• Supplemental vertical load system can be installed to prevent collapse caused by local damage.<br>• The potential for falling debris from taller buildings can be minimized by adding supports and ties on adjacent building, and failure of the roof minimized with roof reinforcing. |
| Configuration | Soft/weak stories | • Disproportionate drift is concentrated on the soft story, potentially causing collapse.<br>• A weak story may not be initially soft, but after yielding as a story first, it becomes soft, and displacements will concentrate in those elements already yielded, potentially causing collapse. | • The most straightforward retrofit is to add elements to the soft or weak story to force displacement from the earthquake to be more evenly distributed throughout the building height.<br>• In some cases of soft stories, it is possible to soften other stories to be more evenly matched. |
| | Discontinuous wall/brace | • Shear walls or braces that do not continue to the foundation create forces at the level of discontinuity that must be designed for, including shear forces that must be transferred into the diaphragm and overturning tension and compression that must be resisted from below | • New walls or braces can be added below<br>• The transfer forces can be made acceptable by reinforcement of the diaphragm and/or strengthening of columns below the ends of the wall/brace. |
| | Set back | • A setback often creates a dynamic discontinuity, because the story below is often much stiffer than the story above. This discontinuity can cause larger-than-expected demands on the floor immediately above. | • The floor above the setback can be strengthened to accept and smooth out the dynamic discontinuity. |
| | Plan Irregularity | • Plan irregularities such as L or T shapes often displace the center of mass from the center of lateral rigidity, causing torsion and resulting in high drifts on some elements.<br>• Re-entrant corners often present in these buildings create large demands on floor diaphragms, tending to pull them apart at these negative corner conditions. | • Lateral force-resisting elements can be added to balance mass and resistance.<br>• Chords and collectors can be added in diaphragms to resist re-entrant corner forces. |

Table 8-3: Common deficiencies by category of buildings 2

| Category | Physical Characteristic | Performance Characteristics | Common Retrofit Techniques |
|---|---|---|---|
| FEMA Model Building Types | W1: Small Wood Frame | • Masonry chimneys normally have incompatible stiffness with the structure and will fail themselves or pull away from the framing. <br>• Cripple stud walls occurring only at the perimeter create a weak/soft story, that often causing the superstructure to topple over. <br>• Discontinuities caused by large garage door openings cause damage. <br>• Hillside structures with weak down-slope lateral anchorage or weak lateral force elements on the down-hill side fail and sometimes slide down the hill. <br>• Buildings with lateral bracing of only stucco or gypsum board can suffer high economic damage. | • Masonry chimneys, reinforced or not, are difficult to make compatible with normally constructed houses. Factory-made light chimneys are the best option. <br>• Most other deficiencies can be mitigated with added elements or added connections to tie the structure together. <br>• Economic damage to gypsum board and hard floor finishes is difficult to control. |
| | W1A Large Wood Frame | • Normally these buildings are more regular than houses, but could suffer similar deficiencies. <br>• The potential infamous deficiency for this building type is the soft/weak story created by ground floor parking—the so-called tuck-under building. Buildings with parking under a concrete first-floor level seldom have the soft story typical of all-wood buildings. | • Mitigate deficiencies similar to W1s. <br>• Add lateral force-resisting elements of wood walls, steel brace frames or moment frames to eliminate the soft/weak story. |
| | W2 Large Wood Post/Beam | • Often create plan irregularity due to plan shape or weakness at lines of the storefronts. | • Add lateral force-resisting elements as required. |
| | S1 Steel Moment Frame | • "Pre-Northridge" welded frames may fracture at the beam-column joint. <br>• Older riveted or bolted frames may also suffer damage at connections <br>• These structures are often very flexible and could collapse from side sway ("P-Delta effect"). <br>• Excessive drift can cause damage to other elements such as interior partitions, stairwells, or cladding. | • Joints can be strengthened. <br>• A diagonal steel bracing or other new lateral force-resisting system can be added. <br>• Dampers can be added to reduce drifts. |
| | S2 Steel Braced Frame | • Braces that are stronger than their connections may fail in the connection, completely losing strength. <br>• Braces can buckle and lose stiffness, causing a significant change in the overall dynamic response. <br>• Certain tube bracing with "thin walls" may fracture and completely lose strength. | • Connections can be strengthened. <br>• Braces can be added or strengthened to lesson effects of buckling. <br>• Thin-walled tubes can be filled with grout or otherwise strengthened to eliminate local wall buckling. <br>• All buildings with a relatively stiff and complete lateral force-resisting system, brittle or not, are candidates for seismic isolation. |
| | S3 Steel Light Frame | • Tension-only braces yield and lengthen, becoming loose. <br>• Braces are often removed. | • Buildings are very light, and deficiencies have seldom caused significant damage. However, braces can be added to add overall strength to building. |

Table 8-3: Common deficiencies by category of buildings 3

| Category | Physical Characteristic | Performance Characteristics | Common Retrofit Techniques |
|---|---|---|---|
| FEMA Model Building Types | S4 Steel w/ Concrete Shear Wall | • This is generally considered a reasonably good building because the shear walls will absorb energy, and the frame will provide reliable gravity load support.<br>• In older buildings, the shear walls may be eccentrically located, causing torsion.<br>• Buildings with exterior concrete pier-and-spandrel enveloping the steel frame may be governed by the concrete response, which may seriously degrade. | • Retrofit measures must be tailored to the specific building and its deficiencies.<br>• Walls can be added to eliminate torsion.<br>• New walls can be added to reduce overall demand on the concrete.<br>• Existing concrete elements can be reinforced to be more ductile.<br>• All buildings with a relatively stiff and complete lateral force-resisting system, brittle or not, are candidates for seismic isolation. |
| | S5 Steel w/Infill Masonry | • This building type is well known for its good performance in the 1906 San Francisco earthquake. There is considerable documentation that shows good performance before the fire.<br>• The brick and steel in this frame are thought to act together to perform well. Cracking and deterioration of the brick, however, may require costly repairs.<br>• Soft stories created by storefronts and entrances may be severely damaged. | • While in a standard evaluation, these buildings may fail the standard, but advanced analysis may show that the composite system requires only minor retrofit.<br>• Shear walls can be added to work with the exterior walls<br>• The walls can be shotcreted from the inside.<br>• Braced steel frames can be added, but the post-buckling response must be considered.<br>• All buildings with a relatively stiff and complete lateral force resisting system, brittle or not, are candidates for seismic isolation. |
| | C1 Concrete Moment Frames | • If designed in accordance with ductile concrete principals (early to mid 1970s, depending on location, this building type will perform well.<br>• If designed and detailed without the special requirements for ductility, this building type could pose serious risk of collapse due to shear failure in the joints or columns and subsequent degradation of strength and stiffness. | • In regions of high seismicity, these buildings are difficult to retrofit by locally improving elements. More likely, a new lateral force system of concrete shear walls or steel braced frames will be required. Added damping will also be effective.<br>• In regions of moderate or low seismicity, local confinement of columns and joint regions might be adequate. |
| | C2 Concrete Shear Walls Type 1 | • A building in which gravity load is carried by the same walls that resist seismic loads is normally considered a higher risk than one with a vertical load-carrying frame (Type 2). However, the bearing-wall building usually has many walls and therefore a low level of demand on the walls. The performance of shear walls is complex and dependent on overturning moment vs. shear capacity ratios. If walls fail in shear, they are likely to degrade, and damage could be major and dangerous.<br>• These buildings often have walls interrupted for large rooms or entrances. These discontinuous walls require special load-transfer details or can cause severe local damage.<br>• Motels and hotels are often built with bearing walls and precast single-span concrete slabs. The bearing of these precast units on the walls and the ties to the walls can be a weakness, particularly if the floors do not have a cast-in-place fill over the precast slab units. | • Discontinuous walls can be mitigated by the introduction of new walls to create continuity, or by adding load-transfer elements at the point of discontinuity.<br>• Wall can be changed from being "shear critical" (failing first in shear - usually bad) to "moment critical" (failing first in moment - usually acceptable) by adding layers of shotcrete or high-strength fiber and epoxy material.<br>• All buildings with a relatively stiff and complete lateral force-resisting system, brittle or not, are candidates for seismic isolation.<br>• In buildings with precast slabs, the bearing of the slabs can be improved by adding a steel or concrete bracket at each wall bearing. Diaphragms can be improved by adding a thin concrete fill or by "stitching" the joints with steel or fiber-epoxy material from beneath. |

Table 8-3: Common deficiences by category of buildings  4

| Category | Physical Characteristic | Performance Characteristics | Common Retrofit Techniques |
|---|---|---|---|
| EMA Model Building Types | C2 Concrete Shear Walls Type 2 | • This building type could also have discontinuous walls similar to Shear Wall Type 1 buildings, with the same result.<br>• These buildings generally have fewer walls than Type 1 and may be understrength, suffering damage to all the walls.<br>• If the walls are not located symmetrically, torsion could result. | • Similar to Shear Wall Type 1, discontinuous shears can be locally retrofit.<br>• Walls can be added to resist torsion and reduce the overall demand on all walls.<br>• Individual walls can be changed from shear critical to moment critical.<br>• All buildings with a relatively stiff and complete lateral force-resisting system, brittle or not, are candidates for seismic isolation. |
| | C3 Concrete w/ Infill Masonry | • These buildings are similar to Type S5, except that their concrete columns are more likely to fail the interaction with the masonry than the steel columns in building type S5 | • See Building Type S5. |
| | PC1 Tilt-Up Concrete Walls | • The classic tilt-up failure consists of the exterior concrete walls pulling away from the roof structure, sometimes causing local collapse of the roof.<br>• Less likely, but possible, is a shear failure within the panel piers due to large window openings. | • Retrofit measures for tilt-ups are well documented in mandatory retrofit ordinances passed in many jurisdictions in California. These requirements consist of creating an adequate connection between panels and roof structure, includings local connector, and transfers from the individual joists and purlins to the diaphragm. In some cases, diaphragms require strengthening or new braces are introduced in the center of the building to reduce diaphragm stress. |
| | PC2 Precast Concrete Walls | • The issue with most precast construction is the connections. Cast-in-place connections have generally performed well while welded ones have not.<br>• Precast walls have seldom been used in regions of high seismicity until recently, when strict code requirements applied. Older precast wall systems may have problems with connections and may be hazardous, particularly if the walls are load bearing. | • Connections must be reinforced to ensure that the structure does not break apart.<br>• It may be difficult to retrofit connections to provide an adequate seismic system, and new elements may also be required.<br>• All buildings with a relatively stiff and complete lateral force-resisting system, brittle or not, are candidates for seismic isolation. |
| | PC2A Precast Concrete Frames | • Older precast frames only exist in regions of moderate or low seismicity. Adequate precast frame systems for high seismicity were only developed and built starting in about 2000.<br>• Older frame structures may have inadequate connections and ductility, and if the structure starts to break apart, partial or complete collapse could follow. | • Connections must be reinforced to ensure that the structure does not break apart.<br>• New elements such as concrete shear walls or steel braced frames will probably be required. |
| | RM1 Rein. Masonry Bearing Walls — Flexible Diaphragms | These buildings are much like tilt-up buildings, with the main weakness likely to be in the wall-to-floor or roof tie. Failure of these ties could lead to local collapse. | Similar to PC1 Tilt-up. |
| | RM2 Rein Masonry Bearing walls — Rigid diaphragm | This building is similar to C2 Concrete Shear Wall Type 2. | See C2, Type 2. |

EXISTING BUILDINGS–EVALUATION AND RETROFIT

Table 8-3: Cmyommon deficiencies by category of buildings 5

| Category | Physical Characteristic | Performance Characteristics | Common Retrofit Techniques |
|---|---|---|---|
| FEMA Model Building Types | URM Unreinforced Masonry Bearing Walls— Flexible Diaphragm | • The primary and most dangerous failure mode is the separation of the exterior wall from the floor/roof and falling outward. Local collapse of floors may follow. In smaller shaking, parapets often fail and fall outward onto the street or adjacent buildings.<br>• The URM walls also often fail in shear with characteristic X cracking. In long-duration shaking, these walls may degrade and lose both lateral and vertical load-carrying ability. | • Several prescriptive ordinances are available to guide evaluation and retrofit.<br>• Partial retrofit ordinances requiring parapet bracing, and/or new wall ties have also been used.<br>• Typical retrofits are to add shear walls or steel braced frames to reduce demand on the URM walls, and to replace the URM wall-to-floor/roof ties and strength shear transfer at these locations. Sometimes, wood diaphragms also need strengthening that can be done by adding plywood. |
| | URMA URM with rigid diaphragms | • This building is similar to URM but will have a shallow arched masonry floor system with wood or concrete overlay, or a concrete slab floor. Flat arched floors have no proven ability to act as a diaphragm and could lead to failures of exterior walls similar to URM, although the ensuing failure of the floors could be much more dangerous.<br>• Concrete slabs, if tied well to the URM walls, may force high in-plane load into the URM walls and cause shear failures. If URM walls are relatively solid, this building type may not form a dangerous collapse hazard. | • Similar to URM, a complete lateral load system must be created to resist out-of-plane URM wall loads and distribute them, through the floor/roof diaphragm to perpendicular walls. Flat arched floors must, at a minimum, be tied together, and cross-building ties must be installed.<br>• URM walls must be adequate for in-plane forces or new elements added.<br>• All buildings with a relatively stiff and complete lateral force-resisting system, brittle or not, are candidates for seismic isolation. |

### 8.8.2 To Learn More

Applied Technology Council, *Seismic Evaluation and Retrofit of Concrete Buildings*, Report No. SSC 96-01, California Seismic Safety Commission, 1996.

A contemporary of FEMA 356 that features many of the same methods and performance terminology. This document contains a good description of retrofit strategies.

Earthquake Engineering Research Institute *Ad Hoc Committee on Seismic Performance, Expected Seismic Performance of Buildings, SP 10*, Earthquake Engineering Research Institute, 1994.

This document was published slightly before FEMA 356 and thus contains slightly different performance terminology. However, photo examples and extensive description of damage states are contained. In addition, estimates are given for the approximate number of various buildings that would be expected to be in various damage states for different ground motion intensities.

Federal Emergency Management Agency, Region 1, *Safeguarding Your Historic Site*, Boston, MA.

This document contains an extensive bibliography covering renovation and repair of existing buildings.

Freeman, John R. *Earthquake Damage and Earthquake Insurance.* McGraw-Hill Book Company, Inc., 1932, New York.

Extremely interesting from a history standpoint, this book Includes discussion of seismology, geotechnical engineering, structural engineering, codes, and loss estimation, and excellent history and available data on earthquakes up to 1932.

Holmes, William T., *Risk Assessment and Retrofit of Existing Buildings*, Proceedings Twelve World Conference on Earthquake Engineering, Auckland, New Zealand, 2000.

This paper contains a more technically oriented description of the methods of FEMA 356 and strategies for design of retrofit systems.

by Gary McGavin

## 9.1 INTRODUCTION

This chapter builds on the knowledge previously presented and discusses the issue of seismic design for nonstructural components and systems. Initially, the primary purpose of seismic design was the desire to protect life safety. Buildings were designed so that the occupants could safely exit the facility following a damaging earthquake. Damage to buildings has always been allowed by the code, even to the extent that the building might need to be demolished following the event, even if correctly designed according to the seismic code. Until very recent years, with minor exceptions, nonstructural design has been minimally required by the model building codes.

The lack of attention to nonstructural systems and their increasing complexity have resulted in the majority of dollar losses to buildings in recent earthquakes. These losses are the result both of the direct cost of damage repair and of functional disruption while repairs are undertaken. Today, good seismic design requires that both structural and nonstructural design be considered together from the outset of the design process. Figures 9-1 and 9-2 illustrate such a design.

Landers Elementary School was constructed and occupied just prior to the magnitude 7.3 Landers earthquake in 1992. The building was situated just 0.4 mile from approximately 10 feet of horizontal offset along the fault trace and experienced severe shaking. Nonstructural damage was minor and included cracked stucco and dislodged suspended ceiling

Figure 9-1: Successful integration of structural and nonstructural design. Landers Elementary School, designed by Ruhnau-McGavin-Ruhnau Associates, 1990.

Figure 9-2: Limited nonstructural damage at Landers Elementary.

tiles in the multipurpose room. In the magnitude 6.5 Big Bear earthquake, which occurred three hours after the Landers event, at the school a water line broke and a hot water heater restraining strap failed due to the incorrect use of lag bolts (too short and not anchored into the studs). The hot water heater remained upright and functional.

As more building owners recognize the necessity to remain operational following a major event, architects will be called upon to provide designs that go beyond the minimal code requirements for life safety and exiting. As our existing stock of older buildings is seismically retrofitted, the nonstructural components and systems must also be seismically retrofitted to the same level as the structure.

## 9.2 WHAT IS MEANT BY THE TERM "NONSTRUCTURAL"

Nonstructural systems and components within a facility are all those parts of a building that do not lie in the primary load-bearing path of the building and are not part of the seismic resisting system. In general, they are designed to support their own weight, which is then transferred to the primary structural system of the building. The number and complexity of nonstructural systems and components far outnumber the structural components of a building. Figure 9-3 shows the basic structural and nonstructural systems.

While nonstructural components are not intended to contribute to seismic resistance, nature does not always respect this distinction. Rigid

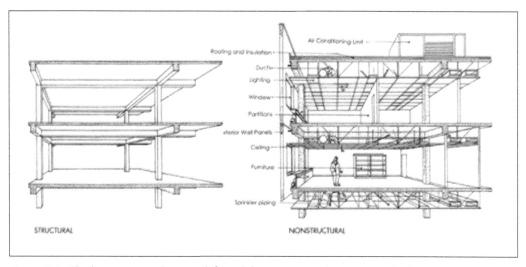

Figure 9-3: The basic structural system (left) and the nonstructural components (right).

nonstructural walls spanning between structural columns will change the local stiffness of the structural system and alter its rsponse, possibly creating a stress concentration. Partitions may suddenly be called upon to perform a supporting role, as seen in Figure 9-4. Conversely, structural members may act in a nonstructural manner if, for example, the contractor omits placing the steel reinforcing in a reinforced concrete wall.

The number and complexity of nonstructural systems and components is very large. A typical broad categorization includes the following:

Figure 9-4: Nonstructural partition walls prevented the total collapse of this unreinforced masonry structure in the 1983 Coalinga earthquake.

❍ Architectural

❍ Electrical

❍ Mechanical

❍ Plumbing

❍ Communications

❍ Contents and Furniture

A more specific list of nonstructural components based on the *International Building Code* (IBC) are categorized as architectural, mechanical, and electrical and is shown below. The IBC also provides design coefficients for each category that are applied to the component to establish the design seismic force.

## 9.2.1 Architectural Components

Interior nonstructural walls and partitions
Cantilever elements
    Parapets
    Chimneys
Exterior nonstructural wall elements and connections
    Light wall elements (metal insulated panels)
    Heavy wall elements (precast concrete)
      Body of panel connections
      Fasteners of the connecting systems
Veneer
    Limited deformability elements
    Low deformability elements
Penthouse (separate from main building structure)
Ceilings
    Suspended
    Attached to rigid sub-frame
Cabinets
    Storage cabinets and laboratory equipment
Access floors
Appendages and ornamentation
Signs and billboards
Other rigid components
Other flexible components

## 9.2.2 Mechanical and Electrical Components

General mechanical
   Boilers and furnaces
   Pressure vessels freestanding and on skirts
   Stacks
   Large cantilevered chimneys
Manufacturing and process machinery
   General
   Conveyors (nonpersonnel)
Piping system
   High deformability elements and attachments
   Limited deformability elements and attachments
   Low deformability elements and attachments
HVAC system equipment
   Vibration isolated
   Nonvibration isolated
   Mounted in-line with ductwork
Elevator components
Escalator components
Trussed towers (freestanding or guyed)
General electrical
   Distributed systems (bus ducts, conduit, cable trays)
   Equipment
Lighting fixtures
   Surface mounted to structure
   Suspended from structure
   Supported by suspended ceiling grid, surface mounted, or hung
   from suspended ceiling

## 9.2.3 Consequences of Inadequate Nonstructural Design

Historically, the seismic performance of nonstructural systems and components has received little attention from designers. The 1971 San Fernando earthquake alerted designers to the issue' mainly because well-designed building structures were able to survive damaging earthquakes while nonstructural components suffered severe damage. It became obvious that much more attention had to be paid to the design of nonstructural components. Some investigators have postulated that nonstructural system or component failure may lead to more injury and death in the future than structural failure.

The following are the basic concerns for nonstructural system/component failure:

○ Direct threat to life

○ Indirect threat to life

○ Loss of building function (loss of revenue and service)

○ High repair costs

## 9.3 NONSTRUCTURAL SEISMIC DESIGN AND "NORMAL" SEISMIC DESIGN

Designing for earthquakes has historically been the domain of the structural engineer. This publication has shown elsewhere what forces are brought to bear on buildings, how the building can be expected to respond due to the earthquake event, what effects the local soils have on the building, and how the building will transmit seismic forces from the foundation up through the structure of the building. By definition, the nonstructural systems and components of the building are attached to the building's primary structure or, in the case of furniture and unsecured equipment/contents, rest unattached on the floors of the building. Seismic forces are generally amplified as they travel up from the foundation through the building to the top of the structure. These increased forces are transmitted to the nonstructural components at their interface with the structure. Many nonstructural systems and components are often very flexible, in contrast to the relatively rigid building structure. This flexibility often leads to a much higher level of excitation than the building's primary structure.

## 9.4 EFFECTS OF IMPROPER NONSTRUCTURAL DESIGN

There are a number of objects that can directly cause either death or injury if they are not properly designed for restraint (Figures 9-5 and 9-6). These injuries are generally due to falling hazards, such as large sections of plaster ceilings, HVAC registers, lights, filing cabinets, etc. There are also indirect threats to life and injury due to nonstructural failures. These might include the inability of occupants to safely exit a building due to damaged materials strewn across the stairs in exit stairwells (Figure 9-6).

NONSTRUCTURAL DESIGN PHILOSOPHY

Although is this example was not a result of an earthquake, the MGM Grand Hotel Las Vegas fire in the 1980s serves to illustrate the complexity of nonstructural design. The fire activated the emergency power supply. The building construction included re-entrant corners that required a large seismic joint between building units. This joint in effect provided a chimney within the building that was not air tight between the floors. The continued operation of the emergency power supply unit caused the asphyxiation of several occupants, since the HVAC system

Figure 9-5: Falling objects can be a direct threat to life, as can be envisioned in this example, had children been sitting in the seats in this elementary school library in Coalinga in 1983.

had its fresh air intake close to the exhaust of the emergency power. Thus, occupants died due to smoke inhalation and carbon monoxide poisoning in part due to the seismic design of the nonstructural systems.

In some cases, nonstructural failure can cause a loss of function of the building. While this may not be critical for some building occupancies, it is very undesirable in others such as hospitals, emergency operations centers, police stations and communications centers. Unfortunately, failures are often caused by system or component interactions. In addition, more and more owners of commercial and industrial facilities are recognizing the need for continued operation in order to reduce financial loss following a damaging earthquake. Hospitals have a need for both continued function and reduction of economic loss. The owner-supplied equipment and contents within a hospital are often significantly more valuable than the building itself. Medications and bandages that are soaked due to flooding from broken fire sprinkler lines cannot be used when they are most needed. The sophisticated equipment within a hospital will take more time to repair, and be more costly, if damaged, than the equipment in an office building or a school.

Figure 9-6: Difficulty in exiting due to debris strewn across exit stairs can be an indirect threat to life, as can be seen in this photo following the Loma Prieta earthquake in 1989.

## 9.5 DAMAGE TO NONSTRUCTURAL SYSTEMS AND COMPONENTS

A lack of attention to detail during the design process is the most likely cause of damage to nonstructural systems and components in a moderate to severe earthquake. This damage poses a threat to the building occupants and may cause the owner significant losses in downtime and repairs. Examples in Table 9-1 illustrate failures in earthquakes that resulted from inadequate nonstructural design.

Nonstructural design philosophy based on the analysis and design of individual components can lead to certain nonstructural failures in moderate and severe earthquakes. The new Olive View Hospital was rendered nonoperational in the Northridge earthquake due to such a philosophy. The building was a replacement for the previous hospital that was so badly damaged structurally in the 1971 earthquake that it needed to be razed. The new replacement hospital was designed as a state-of-the-art facility, and as such, it should have remained operational during the 1994 earthquake. Figure 9-7 shows one of several systems interaction failures that caused the closure of the hospital. The building structural system supported the ceiling system and the fire sprinkler system. The codes require a component approach to seismic qualification for acceptance. The individual components that are analyzed include the ceiling, the lights set in the ceiling, the HVAC system that passes through the ceiling plane, and the main fire sprinkler feed pipe. The Olive View failure occurred when the building structure responded in one manner to the earthquake, the ceiling system and the systems that it supported shook in another manner, and finally the sprinkler system responded in a third

Figure 9-7

Example of systems interaction failure.

NONSTRUCTURAL DESIGN PHILOSOPHY

Table 9-1 Showing Example Damaged Systems/Components and Appropriate Installations

| Building Element | Earthquake Damage | | |
|---|---|---|---|
| **Suspended Ceilings**<br><br>that fall may not be life threatening, but can pose exiting problems for occupants. |  <br>Landers 1992<br>Dropped ceiling below structure. Note no diagonal bracing or compression posts |  <br>Northridge 1994<br>New dropped ceiling below older existing ceiling. Note no diagonal bracing or compression posts. | 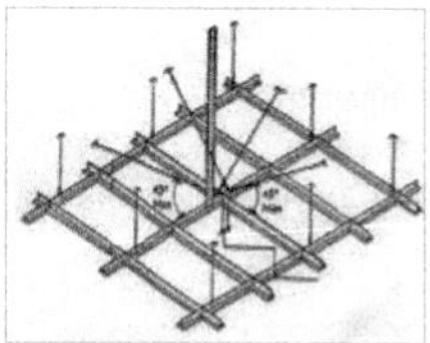 <br>Appropriate Installation. Note diagonal wires and compression posts. Diagonals and compression posts are generally at 144 sq. ft. |
| **Lighting Fixtures**<br><br>can be a direct threat to life, depending on the size of the fixture and the height from which it falls. |  <br>Coalinga 1983 |  <br>Northridge 1994 |  <br>Appropriate Bracing |
| **Doors**<br><br>that fail pose an obvious direct threat to life. Note the fire door in Coalinga 1983 that jammed. |  <br>Photo by Richard Miller<br><br>Santa Barbara 1978 |  <br>Coalinga 1983 |  <br>Northridge 1994 |
| **Windows**<br><br>could pose a direct threat to life, although more often, they are more of a cleanup hazard. |  <br>Coalinga 1983 |  <br>Northridge 1994 |  <br>Hector Mine 1999 |

Table 9-1 Showing Example Damaged Systems/Components and Appropriate Installations (continued)

| Building Element | Earthquake Damage | | |
|---|---|---|---|
| **HVAC Equipment**<br><br>can be a direct threat to life if grills/ducts fall. | <br>Santa Barbara 1978<br>Photo by Richard Miller | <br>Santa Barbara 1978<br>Photo by Richard Miller | <br>Northridge 1994 |
| **Kitchen Equipment**<br><br>can cause a direct threat to life via toppling of equipment and fire/hot liquid burns. | <br>Coalinga 1983 | <br>Coalinga 1983 | <br>Northridge 1994 |
| **Medical Equipment**<br><br>can cause health hazards due to spills. Recalibration is often required. | <br>Northridge 1994 | <br>Northridge 1994 | <br>Northridge 1994 |
| **Emergency Power Supplies**<br><br>have come a long way in the past 30 years, yet they still have difficulty operating following an earthquake. | <br>Joshua Tree 1992 (no failure, base isolated with snubbers) | <br>Northridge 1994 (no failure, hard mounted) | <br>Northridge 1994 (no failure) |

　　　　　　　　　　　　　　　　NONSTRUCTURAL DESIGN PHILOSOPHY

Table 9-1 Showing Example Damaged Systems/Components and Appropriate Installations (continued)

| Building Element | Earthquake Damage | | |
|---|---|---|---|
| **Building Veneer**<br><br>can be a direct threat to life, especially along sidewalks. | <br>Loma Prieta 1989<br>Photo by CA DSA | <br>Hector Mine 1999 | |
| **Elevators**<br><br>should be designed to be operational following an earthquake, but shutdown is required for inspection. | <br>Santa Barbara 1978<br>Photo by Leon Stein | <br>Santa Barbara 1978<br>Photo by Leon Stein | <br>Santa Barbara 1978<br>Photo by Leon Stein |
| **Office Furniture**<br><br>is often owner-supplied and not subject to seismic design by the architect. | <br>Santa Barbara 1978<br>Photo by Richard Miller | <br>Santa Barbara 1978<br>Photo by Richard Miller | <br>Loma Prieta 1989 |
| **Shop Equipment**<br><br>can pose a direct as well as indirect threat to life. | <br>Coalinga 1983 | <br>Northridge 1994 | <br>Northridge 1994 (no failure-properly anchored) |

Table 9-1 Showing Example Damaged Systems/Components and Appropriate Installations (continued)

| Building Element | Earthquake Damage | | |
|---|---|---|---|
| **Piping**<br><br>is especially vulnerable to breakage when it is brittle pipe, and when bending forces are applied to the threads. | <br>Coalinga 1983 | This pipe pounded the wall to its right.<br><br><br>Northridge 1994 | <br>Northridge 1994<br>Brittle pipe failure |
| **Plaster and Stucco**<br><br>seldom will result in a hazard unless it falls from a significant height. | <br>Loma Prieta 1989 | | |
| **Exit Ways**<br><br>may be blocked with debris. | <br>San Fernando 1971<br>Photo by Bill Gates | <br>Loma Prieta 1989 | <br>Northridge 1994 |
| **Hazardous Materials**<br><br>can affect occupants and rescue workers. | <br>Coalinga 1983 | <br>Whittier 1987 | <br>Northridge 1994 |

NONSTRUCTURAL DESIGN PHILOSOPHY

manner. The result was a significant bending moment on the sprinkler drops when the ceiling impacted them, causing cross-thread bending at the joint where the drop connected to the main line. Each of the components was appropriately attached to the building as called for by the code. By not allowing either a flexible joint at the fire line drop or providing a larger hole where the sprinkler penetrates the ceiling plane, the failure of the system was virtually guaranteed.

Many of the failures found in the 1994 Northridge earthquake were a result of systems incompatibilities. It has been long realized that for building structures to survive an earthquake, there must be structural systems compatibility. Few designers would doubt the need for wall systems and roof systems to respond together in an earthquake. Yet, with respect to nonstructural considerations, the interactive nature of these systems has not been fully recognized, and thus, a $1.50 sprinkler pipe failure closed a hospital. Table 9-2 shows examples of failures and success in the design of system interactions.

Systems need to be identified and have a seismic designation and qualification program just as an individual component. Facilities with sophisticated seismic qualification programs such as the Trans-Alaska Pipeline and nuclear power plants have always looked toward qualifying the entire nonstructural systems as well as the individual components.This type of procedure is especially important where specific functions must be maintained, such as with an emergency power supply in a hospital.

Moving equipment such as reciprocating pumps need isolation so as to not interject unwanted vibrations into the building structure. This isolation needs to be "snubbed" in order to limit the lateral excursions of the equipment.

Table 9-2 Examples of Systems Interactions – Failures and Successes

| illustration | description | illustration | description |
|---|---|---|---|
| | Interaction of pendant light with glazing separating spaces in the 1994 Northridge earthquake. The pendant lights swung longitudinally, breaking the glass. Current codes require bracing of pendant lights to prevent swinging in both the transverse and longitudinal direction if they can impact other objects up to an angle of 45 degrees (1 g acceleration). | | The suspended ceiling surrounding the column did not allow for ceiling movement, causing ceiling to fail at the column/ceiling interface. |
| | While this system was not subjected to an earthquake, it will probably suffer the same failure as seen in the photo above. This condition is increased in its complexity because of the fire separation above the suspended ceiling running diagonally from the upper right. | | Vacuum System in the Coalinga Hospital that was operational following the 1983 earthquake. The tank was anchored and the flexible line to the tank prevented damage. |
| | Building primary structure moved in one manner, the substructure (covered walk on the right) moved differentially, causing both the HVAC duct and electrical conduit to fail in the 1994 Northridge earthquake. | | While the damage to the suspended ceiling is most evident in this photo, the exterior nonstructural wall failed on this bowling alley in the 1992 Landers earthquake leaving one side of the building completely open. The failure was due to the large length of the wall that used small fasteners for anchorage to the primary structure (tapered beam). |
| | Building seismic separation that performed successfully in the 1978 Santa Barbara earthquake. The cosmetic trim panel was damaged, as expected.<br>Photo by Richard Miller | | Bringing large utility lines into a building that is base isolated requires consideration of systems interaction. Here, the water line has a braided section to allow for differential movement between the building and the utility line. |

NONSTRUCTURAL DESIGN PHILOSOPHY

| illustration | description | illustration | description |
|---|---|---|---|
| | This ceiling displaced and caused the water supply line that passed through the ceiling to bend and break the threads in the pipe above the ceiling during the earthquake. Water leaked for some time into the ceiling cavity following the 1994 Northridge earthquake, finally causing a collapse of the hard ceiling and the glue-on panels, months after the earthquake. | | The rectangular building configuration of this library had stiff side walls and a relatively stiff rear wall. The large expanse of glass on the front wall allowed for excessive torsional movements in the 1983 Coalinga earthquake. The larger upper story window panes broke. As a general rule, smaller window panes perform better in earthquakes. |
| | The diagonal member running from the top left was a horizontal, nonstructural architectural appendage that separated the buildings on the left from the buildings on the right. When the garage spaces on each building failed in the 1994 Northridge earthquake, they in turn caused the failure by collapse of the architectural appendage and blockage of the access between the buildings. Emergency vehicles were unable to enter the space between the buildings due to the collapse of this nonstructural appendage. | | Ceiling tiled popped out of this suspended ceiling during the 1992 Landers earthquake due to the high bay, large suspension length below the structure and a lack of compression posts to prevent the ceiling from lifting in a wave like fashion during the earthquake. Adding to the failure of this ceiling was poor workmanship at the ceiling grid joints. |
| | | | This cast-iron brittle pipe entered the utility room through the stiff concrete slab on the right and exited the room through a one-hour fire wall to the left. The rigid nature of the two connections on either side of the valve caused its failure in the 1994 Northridge earthquake. |

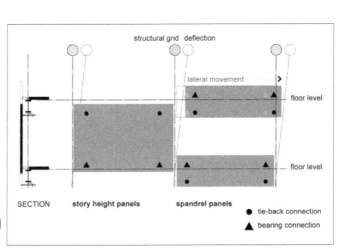

## 9.6 DESIGN DETAILS FOR NONSTRUCTURAL DAMAGE REDUCTION

bearing connection

flexible or tie-back connection

bearing connection

flexible or tie-back connection

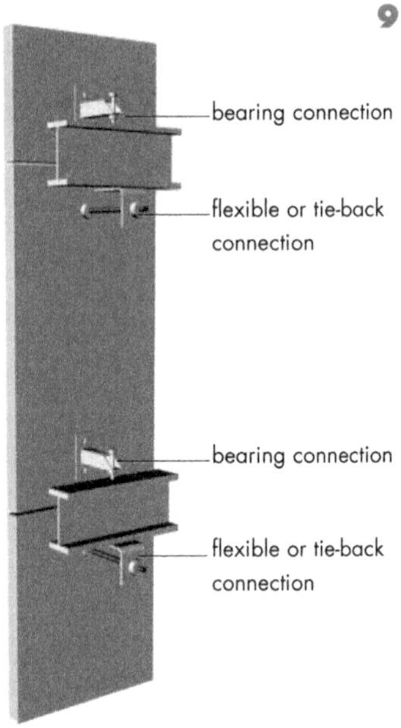

Figure 9-8: Typical push-pull connection for precast panel.

This section shows some examples of conceptual details for a number of typical nonstructural components.  These are intended to give an indication of appropriate design approaches and should not be used as construction documentation.  It will be seen that these approaches mostly consist of providing adequate support and  supplementary lateral bracing, or isolating nonstructural components from the building structure to reduce undesirable interaction between nonstructural and structural elements.  The isolation issue is discussed in more detail in Section 9.7

### 9.6.1 Precast Concrete Cladding Panels

Figure 9-8 shows a typical "push-pull" connection for a precast panel that spans between floors.  The bottom connection provides bearing; the top connection uses a steel rod that is designed to bend under lateral drift.  The rod must be strong enough, how-ever, to resist out-of-plane wind loads.  The bearing connections may also be located at the top and the flexible connection at the bottom.

Figure 9-9 shows a typical layout of the supports for a story height panel and spandrel panels, and diagrams the lateral structural movement that must be accommodated.

structural grid  deflection

lateral movement

floor level

floor level

SECTION    story height panels    spandrel panels    ● tie-back connection
▲ bearing connection

Figure 9-9:  Connection types and locations for precast panels.

NONSTRUCTURAL DESIGN PHILOSOPHY

### 9.6.2 Suspended Ceilings

Figure 9-10 shows typical suspended-ceiling bracing. Diagonal bracing by wires or rigid members: spacing should not be more than 144 sq. ft. The vertical strut is recommended for large ceiling areas in high seismic zones; it may be provided by a piece of metal conduit or angle section.

### 9.6.3 Lighting Fixtures

Heavy fluorescent light fixtures inserted in suspended ceilings must be supported independently, so that if the grid fails, the fixture will not fall. Figure 9-11 shows a lighting fixture with two safety wires located at the diagonal. For heavy fixtures, four wires must be provided. Suspended fixtures must be free to swing without hitting adjoining components.

### 9.6.4 Heavy (Masonry) Full-Height Non load Bearing Walls

Heavy partitions, such as concrete block, should be separated from surrounding structure to avoid local stiffening of the structure and to avoid transmitting racking forces into the wall. Figure 9-12 shows two approaches for providing sliding or ductile connections at the head of full-story masonry partitions.

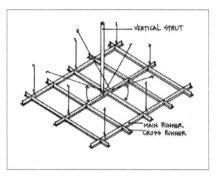

Figure 9-10: Suspended-ceiling seismic bracing.

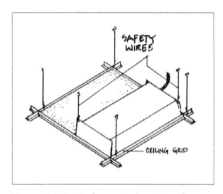

Figure 9-11: Safety wire locations for fixture supported by suspended-ceiling grid.

Figure 9-12: Attachment for full-height masonry partition wall that allows relative longitudinal movement (EQE).

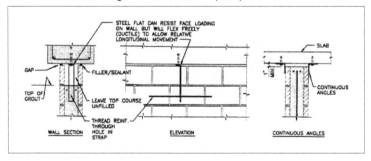

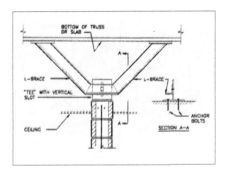

Figure 9-13: Seismic bracing for masonry partial height partition wall (EQE).

### 9.6.5 Partial-Height Masonry Walls

Figure 9-13 shows an overhead bracing system for a partial height wall. The bracing used should have some degree of lateral flexibility so that structural deflections parallel to the wall do not transmit forces into the system. Where vertical deflections due to dead loads, live loads, and seismic forces could occur, a slotted hole as shown, or some similar provision, should be made to prevent vertical loading of the wall.

### 9.6.6 Partial-Height Metal Stud Walls

Metal stud partitions that terminate at a suspended ceiling should be braced independently to the building structure, as shown in Figure 9-14. Normal office height partitions can be braced by a single diagonal angle or stud brace.

### 9.6.7 Parapet Bracing

Heavy parapets should be braced back to the roof structure. This is a typical problem with unreinforced masonry buildings, which often have large unsupported parapets. Figure 9-15 shows bracing for an existing masonry parapet; the roof structure should also be securely tied to the wall (not shown).

Figure 9-14: Bracing for partial height stud wall (EQE).

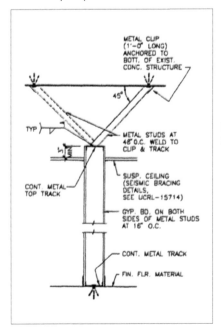

Figure 9-15: Bracing for existing unreinforced masonry parapet.

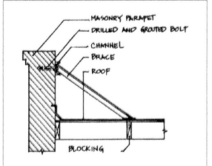

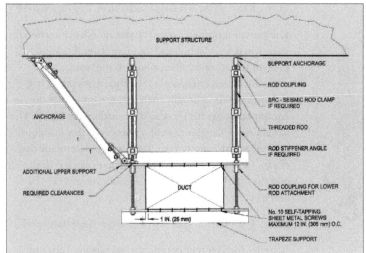

Figure 9-16:

Typical duct bracing system (Mason Industries Inc).

## 9.6.8 Sheet Metal Ductwork

Figure 9-16 shows a typical support and bracing system for large duct-work in a high seismic zone. The seismic code specifies the size of ducts and length of support that require seismic bracing.

## 9.6.9 Piping

Figure 9-17 shows typical bracing for large diameter piping. The seismic code specifies the types and diameter of piping, and length and type of hanger, that require seismic bracing.

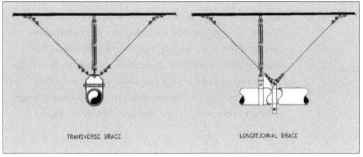

Figure 9-17:

Typical bracing for piping (Mason Industries Inc).

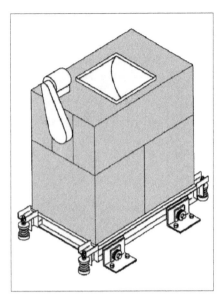

Figure 9-18: Vibration-isolated chiller with snubbers to restrict lateral movement (Mason Industries Inc).

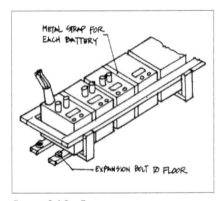

Figure 9-19: Emergency power battery rack support.

### 9.6.10  Vibration-Isolated Equipment

Equipment mounted on spring vibration isolators needs to be fitted with "snubbers" that limit lateral motion to prevent the equipment toppling off the isolators and suffering damage (Figure 9-18). The frequency of the isolation system is usually such that the motion of the equipment is greatly increased by earthquake forces. The snubbers are faced with resilient material that cushions any impacts that may occur. Detailed guidelines for the design of seismic restraints for mechanical, electrical, and duct and pipe are in FEMA publications 412, 413 and 414.

### 9.6.11  Emergency Power Equipment

Batteries for emergency power need positive restraint. Figure 9-19 shows a custom designed rack, constructed from steel sections, to support and restrain a set of batteries. The batteries are also strapped to the rack for positive restraint. Alternative emergency power sources, such as gas or oil, need flexible utility connections and restrained equipment.

### 9.6.12  Tall Shelving

Tall shelves, such as library shelves, are often heavily loaded and acceleration sensitive. They need longitudinal bracing and attachment to the floor. The top bracing should be attached to the building structure and strong enough to resist buckling when the heavy shelves attempt to overturn (Figure 9-20).

### 9.6.13  Gas Water Heaters

Gas water heaters need restraint to prevent the heater tank from toppling and breaking the gas connection, causing a fire risk. Figure 9-21 shows a domestic hot water heater installation. A flexible gas connection is desirable but not essential if the tank is well restrained. The bottom restraint can be provided by an additional strap, or by securely bolting the base support to the floor.

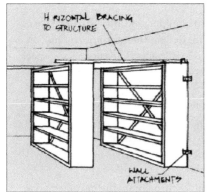

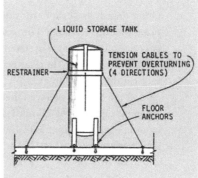

Figure 9-20: Typical layout of bracing for tall shelving.

Figure 9-21: Anchorage of free-standing upright liquid storage tank.
SOURCE: EQE.

## 9.7 THE NEED FOR SYSTEMS DESIGN

Following a damaging earthquake, whether large or small, the public expects to be able continue to use many building types. The building types that have the mandate for some post-earthquake operation tend to be essential facilities, such as acute care hospitals and those buildings where the owners see a clear financial benefit for continued operation. Unfortunately, the expectations of the performance of our essential facilities are often not realized, and a number of modern healthcare facilities close with all too much regularity following even moderate earthquakes.

Over the years, codes have become more and more sophisticated with respect to structural seismic integrity. Unfortunately, there remains a lack of understanding by many with respect to building function, in which the nonstructural systems play the key role. The philosophy of code implementation carried out via the model codes, including the *International Building Code* (IBC), is based on the seismic provisions developed in the National Earthquake Hazard Reduction Program (NEHRP) and the *Uniform Building Code* (UBC), which uses the Structural Engineers Association of California recommendations. Both require simple component anchorage, which does not address function to the necessary level, especially for essential facilities.

Other design professions outside the building industry have long recognized that there is a need for systems design when continued function is necessary. The practice of systems design can be witnessed in aerospace design, naval design, nuclear power design, weapons design, and even race car design. As an example, few would accept rides on modern aircraft if it was not believed that the aircraft had been designed from a systems engineering point of view. It would not be acceptable for the landing gear to pinch the hydraulic brake lines when the wheels fold into the fuselage.

Systems failures in health care facilities throughout the impacted area in the Northridge earthquake prompted the California Seismic Safety Commission to sponsor Senator Alfred Alquist's Hospital Seismic Safety Act of 1994. This legislation was clear in its direction to the industry and design professionals to maintain the operability of health care facilities following future damaging earthquakes. The legislation called for consideration of systems design. Hospitals and other essential facilities complying with the legislation will have a lower risk of failure in future earthquakes.

The California Hospital Seismic Safety Act of 1994 took the further step of identifying building contents within the design parameters. Prior to the implementation of this legislation, building contents had been left out of the qualification procedures in almost all codes, unless they met certain criteria. Without consideration of building contents, hospitals continue to be more vulnerable to failure due to earthquake shaking.

## 9.8 WHO IS RESPONSIBLE FOR DESIGN?

There is no clear answer for the responsibility of many nonstructural design issues. In order to help designers determine who is responsible for nonstructural issues for both systems and components, Table 9-3 is provided as a guide. Architects may wish to use this as a guide in establishing contractual relationships with their consultants prior to beginning design. It is certainly helpful to all design professionals to know who is responsible for specific tasks. It should be noted that there are many cases where design responsibilities are not clear, even when a responsibility chart such as that below is used. The architect, as the design professional in charge, must ensure that the assigned responsibilities are clearly defined.

Table 9-3 Design Responsibilities for Nonstructural Components

The following list can be reviewed and modified by architects for their specific project. The table is not intended to apply to every project, but rather to act as a check list and a guide.

| Nonstructual System or Component | Architect | Structural Engineer | Electrical Engineer | Mechanical Engineer | other design professionals | Remarks |
|---|---|---|---|---|---|---|
| curtain wall | 1 | 2 | | | consider a specialty consultant | Small glazing panes perform better in earthquakes. Avoid window film unless properly applied. |
| doors / windows | 1 | | | | | Consider how doors will avoid racking in nonstructural walls. |
| access floors | 1 | | | | consider a specialty consultant | Consider in-the-floor ducts rather than raised floors where practical |
| HVAC systems | 2 | | | 1 | | Systems that require vibration isolation also require snubbing. |
| plumbing systems | 2 | | | 2 | | Vertical plumbing runs are subject to floor to floor drift |
| communication systems | 2 | | 1 | | 1 consider a specialty consultant | Some communications systems come as a package. Make sure that they interface with the building appropriately. |
| data systems | 2 | | 1 | | 1 consider a specialty consultant | Consider support systems such as cooling environments. |
| elevator systems | 1 | 2 | 2 | 2 | 2 | Design some elevators to operate after the earthquake |
| emergency power supply system | 2 | 2 | 1 | 2 | 2 | All systems interfaces need to be considered as their vulnerability can cause an entire facility to become non-operational. |
| fire protection systems | 2 | | 2 | 1 | 1 consider a specialty consultant | Floor to floor piping is subject to story drift. |
| kitchen systems | 1 | | | | 2 consider a specialty consultant | |
| lighting systems | 2 | | 1 | | | |

Table 9-3 Design Responsibilities for Nonstructural Components (continued)

| Nonstructual System or Component | Architect | Structural Engineer | Electrical Engineer | Mechanical Engineer | other design professionals | Remarks |
|---|---|---|---|---|---|---|
| medical sytems | 1 | 2 | 2 | 2 | 1 consider a specialty consultant | Often, the architect needs to provide protection to equipment as it is outside the code requirements. |
| ceiling systems | 1 | 2 | 2 | 2 | | Avoid drop ceiling elevation changes. Avoid Large ceiling cavities. |
| unbraced walls and parapets | 1 | 2 | | | | |
| interior bearing walls | 1 | 2 | | | | |
| interior non-bearing walls | 1 | | | | | Consider earthquake effects on doors for egress. |
| prefabricated elements (architectural appendages) | 1 | 2 | | | | |
| chimneys | 1 | 2 | | | | |
| signs | 1 | 2 | | | | |
| billboards | 2 | 1 | 2 | | 2 consider a specialty consultant | |
| storage racks | 1 | | | | 2 consider a specialty consultant | Proprietary manufactured racks may or may not include seismic design considerations. |
| cabinets and book stacks | 1 | | | | 2 | Architect needs to provide proper wall backing. |
| wall hung cabinets | 1 | | | 1 | | Architect needs to provide proper wall backing. |
| tanks and vessels | 2 | 2 | | | | |
| electrical equipment | 2 | 2 | 1 | | | |
| plumbing equipment | 2 | 2 | | 1 | | |

Note:     1 = Primary Responsibility     2 = Support Responsibility

NONSTRUCTURAL DESIGN PHILOSOPHY

## 9.9 NONSTRUCTURAL CODES

The early 1970s saw the first inclusion for nonstructural provisions other than walls, parapets, and chimneys. The provisions have grown to include a wide variety of nonstructural components and building systems since the mid-1970s, but the seismic codes have yet to recognize the need for qualification of owner-supplied equipment that is not fixed to the building. The codes have also yet to come to grips with systems qualification and continued performance for facilities. A discussion of the philosophy of codes for nonstructural components and systems is presented in Chapter 6, Section 6.6.

## 9.10 METHODS OF SEISMIC QUALIFICATION

Qualification involves the acceptance of components and systems for use in a seismic environment and compliance with code requirements. There are numerous methods by which seismic qualification can be realized. Each method has a narrow window of applications for effective seismic qualification. These are:

○ Design Team Judgment

○ Prior Qualification

○ Physical Tests

○ Mathematical Analysis

○ Static Equivalent Analysis

○ Dynamic Analysis

### 9.10.1 Design Team Judgment

Design team judgment is a valuable resource in any seismic qualification program. An inappropriate selection of equipment or qualification method by the design team may lead to nonstructural system failures during an earthquake. The design team needs to meet early in the programming, schematic, and design development phases to discuss the various systems to be used in a facility design. Just as in preliminary design the architect is interested in how deep structural members need to be for floor spans, and how much room is needed above the finished ceiling and below the structural system for HVAC and lighting, the architect and engineers need to discuss how movement of the various systems will impact each other in a sizeable earthquake. Early appropriate building configuration decisions by the design team will have a great impact on the success of a building structure in an earthquake. Similar appropriate decisions by the design team early in the design process will have a profound impact on the design of nonstructural systems and components. Following the early discussions of the design team, more quantitative methods of qualification can be utilized, as discussed in the following sections.

## 9.10.2 Prior Qualification

In some cases, a product or system can be shown to have a previous qualification procedure that exceeds the code requirements. This might include equipment that has been designed for shipboard installation, where the accelerations expected and durations of those accelerations far exceed the code requirements of a simple static mathematical model. Such equipment is arguably considered to be qualified by prior experience. Other examples might include a component manufacturer that has had their equipment tested on seismic shaking equipment that provides a stringent-enough test to envelope the specific installation, providing a prior qualification. When available, the manufacturers will provide test results and/or reports detailing the testing program for the building official's review.

Appropriate detailing employing prior qualification is often used by the architect. Consider, for example, suspended ceiling systems. Most architects use suspended ceiling details that have proven over time to be effective, either through prior seismic experience or prior analysis. The familiar 45° splay wires at areas of interval, such as 144 sq. ft., are such an example. There is no need for the architect to recalculate the forces in the resisting wires for each new application. Choosing the appropriate detail is often sufficient. There are, however, limits on "standard details", and the architect should always review each detail to be used for the specific application.

## 9.10.3 Mathematical Analysis and Other Qualification Methods

There are two basic forms of analysis. The first is a simple static equivalent analysis. This is the method most suited for most simple architectural problems for nonstructural design. The second is the more complex dynamic analysis which requires costly engineering analysis and is seldom used for nonstructural components. In place of, or in combination with, physical testing the latter may be used.

These methods of qualification are generally not going to be employed by architects in "normal" designs. In general, they are better suited for product/system manufacturers and researchers that want to show either a wide range of possible applications or to confirm predicted responses. All of these qualification methods are expensive and time consuming.

## 9.11 SOME MYTHS REGARDING NONSTRUCTURAL DESIGN

● "My Engineers take care of all my seismic design"

Many architects believe that seismic design is controlled in total by their engineers and they should not be involved in the conceptualization or the coordination of seismic design. As discussed throughout this work by the various authors, successful seismic design begins and ends with the architect. It is true that the engineers may control the details of many components within the facility, but it is the architect who must understand the interrelationship between the various systems within the facility for successful performance during and after an earthquake.

● "My building is base isolated ... I don't need to worry about the nonstructural components"

Building base isolation in general reduces the effects of horizontal motions within a building, but it does not eliminate them. The architect and design team should be aware of the limitations of base isolation and special conditions, such as the need for utilities to accommodate large lateral movement where they enter the building above or below ground.

● "Window films protect windows from breakage in an earthquake"

If properly applied, window films can reduce some glass breakage. The film needs to be taken all the way to the edge of the glass. Often film to be applied to the glazing, while it is mounted in its frame, and is then cut with a razor blade against the mullions and muttons that score the glass, making it vulnerable to breakage during violent shaking. As the

Figure 9-22:  This lathe lifted up in the air and came down about four inches from its starting point before sliding another six inches, where it finally came to rest in the 1983 Coalinga earthquake.  The initial complete lifting off the floor is evidence of vertical acceleration of more than 1g.

film ages, it can lose its adhesion characteristics.  If films are used, they should be cross-ply films and have undergone aging tests to predict how they will outgas and to what extent they will become brittle.

● "My building in San Bernardino survived the 1994 Northridge earthquake … it is earthquake proof"

San Bernardino (or any other appropriate location) was a long distance from the 1994 Northridge earthquake.  While distance does not always guarantee safety (San Francisco was approximately 60 miles from the focus of the Loma Prieta earthquake), in general being a substantial distance from the earthquake will lessen the effects of the earthquake on the building and its nonstructural components.  Nonstructural failures are commonly seen at greater distances than structural failures.  This is especially the case where the building components are not designed for the earthquake environment.

- "Vertical motions in earthquakes do not need to be considered for nonstructural design"

Historically, the model earthquake codes paid little attention to the vertical component of shaking generated by earthquakes. As a rule of thumb, the maximum vertical ground motion is generally 60 to 70% of the maximum horizontal ground motion. While it may be unnecessary to consider the vertical motions of the structure as a whole, this is often not the case with nonstructural design. The model codes have little reference to vertical acceleration design requirements for nonstructural components. The building, usually due to its architectural configuration, can act as an amplifier for both horizontal and vertical motions. Therefore, even though the code most often does not require vertical design resistance, the designer must be cognizant of the implications of vertical motions during an earthquake and their potential effects (Figure 9-22).

## 9.12 WHAT CAN THE ARCHITECT DO TO DECREASE NONSTRUCTURAL DAMAGE?

Not only does the architect have the obligation to coordinate the overall design of the building, but the architect is also responsible for the basic seismic safety of the design.

The architect should guide the other design professionals in the design decisions, rather than simply turning the design over to the project engineers. Since many nonstructural issues involve the intermixing of several engineering professions, the architect should understand how each system will react with the other building systems. The architect needs to be able to visualize the system, its components, and how they will interact in an earthquake, strong winds, fire, etc. The architect should sit down with the consulting engineers early and often, beginning with a discussion of the earthquake performance objectives for the facility, to permit each of the disciplines to see the potential for interactions between systems and components. Office standards should be developed for the interfacing of systems and components.

Simple designs make design life easier. The current vogue for complex shapes in architectural design increase the complexity of the nonstructural systems. This increase in complexity decreases the architect's ability to visualize how systems and components will respond and interact.

## 9.13 THE COMPLEXITY OF RETROFITTING EXISTING BUILDINGS

The nonstructural implications of retrofitting existing buildings can be very complex. In many cases, the structural seismic retrofit may be compromised due to historic codes or other considerations. In some cases, buildings have grown over time, which often means that nonstructural systems pass through more than one time era of construction. These interfaces may make the nonstructural retrofit very difficult. California architects and engineers are faced with a particular difficulty in this issue, based on the 1994 law requiring the upgrading of all existing acute care hospitals by 2030 (SB1953). In fact, the task may be essentially impossible, and following the expense of the retrofits, hospitals may yet run a high risk of seismic failure due to nonstructural systems/component failures.

A better solution than expecting performance out of systems that are difficult if not impossible to retrofit might be a "hospital lifeboat". The lifeboats could be self-contained, factory-built modular buildings sized to accommodate the expected emergency population needing attention following a major earthquake. The lifeboats would be permanently stationed on the health care campus, ready for operation as needed. These lifeboats could be provided at a fraction of the cost of the thorough retrofits currently required by SB1953, saving California money and providing a much higher degree of confidence that the hospitals will be operational for emergency services following an earthquake.

For facility types other than acute care facilities, the design team needs to identify how the existing building will react to the new structural improvements, and how these will impact the nonstructural elements of the facility. The design professional may be required to determine which systems/components can fail and simply protect the occupants from falling hazards.

## 9.14 CONCLUSIONS

The largest immediate strides in resisting the impacts of nonstructural failures in future earthquakes will come from designers who understand the implications of systems design. Next, there will be great increases in nonstructural seismic resistance by designers implementing the keep-it-simple philosophy. Apart from that, there will be great strides in structural systems that will result in reduced-motion inputs at the non-structural system/component interfaces. These include base isolation as it is currently being employed. Another future structural improvement that will likely improve nonstructural seismic performance is the increasing use of both active and passive seismic dampers. These systems show promise for significantly decreasing building motions that in turn will decrease nonstructural damage. Some active dampers under development will be able to respond to the earthquake in almost real time.

On the distant horizon is the transformation of the building structure from its seemingly rigid skeleton to a skeleton with a muscle system similar to an organism. Shape-memory materials such as nitinol can be fabricated to act like muscles. The nitinol reduces its size when heated, rather than expanding like most construction materials. These "building muscles" can then be either tightened or relaxed with electrical input (heat), so that the building achieves "balance" during a seismic event in much the same way that our bodies can remain balanced when we stand on a moving bus or train. As these shape-memory materials reduce the effects of acceleration on the building as a whole, they will reduce the large acceleration inputs on the nonstructural systems/components. Shape-memory construction has been successfully utilized in aerospace design in recent years. It will take considerable time for it to be successfully used in building design, although some limited research has been reported.

## 9.15 REFERENCES

2001. *California Building Code* based on the 1997 *Uniform Building Code*, International Conference of Building Officials, Whittier, CA.

2003 *International Building Code*, International Code Council, Country Club Hills, IL

FEMA 74, *Reducing Risks of Nonstructural Earthquake Damage*, FEMA, Washington D.C. 1994.

EQE, *Structural Concepts and Details for Seismic Design*, Lawrence Livermore National Laboratory, Livermore, CA. 1991.

Mason Industries Inc., *Seismic Restraint Guidelines*, Mason Industries Inc, Hauppauge, NY. 1999.

By Christopher Arnold

## 10.1 INTRODUCTION

Earthquakes are only one of several hazards to which buildings are vulnerable. The two other significant natural hazards are floods and high winds, including tornadoes. These hazards are extreme variants of benign natural processes. Earthquakes represent a highly accelerated instance of the slow adjustment that the earth makes as it cools. High winds and tornadoes are an exaggerated form of the pleasant winds and breezes that freshen our everyday existence. Devastating floods are the result of excessive localized rainfall that, when normal, is necessary for the provision of water supply and the nurturing of plant life. These natural hazards are not aberrations, are not malicious, and are part of nature's order.

The traditional hazard to which buildings are vulnerable is fire, and history abounds with urban fire disasters right up to the present day. Fire disasters are usually due to human errors and carelessness, but are sometimes originated by earthquakes and made more lethal by winds (Figure 10-1). Fire, however, when properly controlled, is an important contrib-

Figure 10-1

Earthquake and Fire, Marina District, Loma Prieta, 1989.

SOURCE: EERI. JOHN EGAN/ GEOMATRIX CONSULTANTS

utor to human comfort. Sometimes, of course, a fire disaster is the result of malicious intent. The newest hazard - that of a terrorist attack on a building - is malicious, but is also an extreme form of everyday circumstance that we have come to accept: criminality. Building design has long recognized the need for locks and, more recently, remote cameras and sensing devices are also designed to prevent criminal - and now terrorists - from gaining entry to our buildings[1]

## 10.2 MULTIHAZARD DESIGN SYSTEM INTERACTIONS

This publication focuses on design against earthquakes, but the other hazards must also be assessed. Each of them has differing levels of risk—i.e., the probabilities and consequences of an event. Some, such as earthquakes, floods, and high winds, are specific to certain regions. The risks of terrorism are still uncertain compared to those of natural hazards that have a long history of statistical and scientific observation and analysis. Fires are more pervasive than any of the natural hazards. However, design against fire has long been built into our building codes, in the form of approved materials, fire-resistant assemblies, exiting requirements, the width and design of stairs, the dimensions of corridors, and many other issues.

An important aspect of designing against a single hazard such as earthquakes is the extent to which the design methods may reinforce or conflict with those necessary for protection against other hazards. Multihazard design involves a risk assessment of all hazards at a programmatic stage to ensure that protection measures are not in conflict. Ideally, the measures used would focus on reinforcement rather than conflict, so that the overall risk management plan enables the cost of construction to be reduced.

To assist the reader in evaluating the interactions between protective design methods, Table 10-1 summarizes the effects that seismic design measures may have on performance of the building in relation to other hazards.

---

[1] More buildings have been destroyed by war in the 20th century than by all natural disasters; modern terrorism is the latest variation on traditional wartime urban destruction by shells or bombs.

The horizontal rows show the five primary hazards. The vertical rows show methods of protection for the building systems and components that have significant interaction, either reinforcement or conflict. These methods are based on commonly accepted methods of risk reduction for the three main natural hazards, together with fire protection methods, and the methods for security/blast protection presented in FEMA 426, *Reference Manual to Mitigate Potential Terrorist Attacks against Buildings*, and FEMA 430, *Site Design Guidance to Mitigate Potential Terrorist Attacks*.

The comments in this matrix are not absolute restrictions or recommendations, but rather are intended to provoke thought and further design integration. Reinforcement between hazards may be gained, and undesirable conditions and conflicts can be resolved by coordinated design between the consultants, starting at the inception of design.

Table 10-1 provides information to help the reader develop a list of reinforcements and conflicts for the particular combination of hazards that may be faced. Development of lists such as these can be used to structure initial discussions on the impact of multi-hazard design on the building performance and cost that, in turn, guide an integrated design strategy for protection. The system and component heading list is similar to that used for the building security assessment checklist in FEMA 426, Reference Manual to Mitigate Potential Terrorist Attacks against Buildings.

The following Table, Multi-hazard Design System Interactions, refers to the typical structures illustrated in Chapter 8. An explanation of the symbols used is below:

**+** Indicates desirable condition or method for designated component/system (cell color green)

**-** Indicates undesirable condition for designated component/system (cell color red)

**o** Indicates little or no significance for designated component/system (cell color yellow)

**+/-** Indicates significance may vary, see discussion column

Table 10-1 Multihazard Design System Interactions

| | | The Hazards | | | | | |
|---|---|---|---|---|---|---|---|
| System ID | Existing Conditions or Proposed Protection Methods | Earthquakes | Flood | Wind | Security/Blast (FEMA 426) | Fire | Discussion Issues |
| **Building System Protection Methods: Reinforcements and Conflicts** | | | | | | | |
| **1** | **Site** | | | | | | |
| | 1-1 Site-specific hazard analysis | + | + | + | + | + | Beneficial for all hazards |
| | 1-2 Two or more means of site access | + | + | + | + | + | Beneficial for all hazards |
| **2** | **Architectural** | | | | | | |
| **2A** | **Configuration** | | | | | | |
| | 2A-1 Reentrant-corner plan forms | - | o | - | - | o | May cause stress concentrations and torsion in earthquakes, and concentrate wind and blast forces. |
| | 2A-2 Enclosed-courtyard building forms | - | o | + | +/- | o | May cause stress concentrations and torsion in earthquakes; courtyard provides protected area against high winds. Depending on individual design, they may offer protection or be undesirable during a blast event. If they are not enclosed on all four sides, the "U" shape or reentrant corners create blast vulnerability. If enclosed on all sides, they might experience significant blast pressures, depending on roof and building design. Since most courtyards have significant glazed areas, they could be problematic |
| | 2A-3 Very complex building forms | - | - | - | - | - | May cause stress concentrations, torsion, and indirect load paths in highly stressed structures, and confusing evacuation paths and access for firefighting. Complicates flood resistance by means other than fill. |
| | 2-A4 Large roof overhangs | - | o | - | - | o | Possibly vulnerable to vertical earthquake forces. Wall-to-roof intersection will tend to contain and concentrate blast forces, if the point of detonation is below the eaves. |
| **2B** | **Planning and Function** | | | | | | |
| | not applicable | | | | | | |

DESIGN FOR EXTREME HAZARDS

Table 10-1 Multihazard Design System Interactions (continued)

| System ID | Existing Conditions or Proposed Protection Methods | Earthquakes | Flood | Wind | Security/Blast (FEMA 426) | Fire | Discussion Issues |
|---|---|---|---|---|---|---|---|
| | **Building System Protection Methods: Reinforcements and Conflicts** | | | | | | |
| | **The Hazards** | | | | | | |
| **2C** | **Ceilings** | | | | | | |
| | 2C-1 Hung ceiling diagonally braced to structure | + | o | + | + | + | Reduced damage from earthquake, wind forces, blast. If part of fire protection system, increases possibility of retaining integrity. |
| **2D** | **Partitions** | | | | | | |
| | 2D-1 Concrete block, hollow clay tile partitions | - | + | - | - | + | Earthquake and wind force reactions similar to heavy unreinforced wall sections, with risk of overturning. Tile may become flying debris in blast. It is possible but difficult to protect structures with blast walls, but a weak nonstructural wall has more chance of hurting people as debris. Desirable against fire and not seriously damaged by flood. |
| | 2D-2 Use of nonrigid (ductile) connections for attachment of interior non load-bearing walls to structure | + | o | + | + | - | Non rigid connections necessary to avoid partitions that influence structural response. However, gaps provided for this threaten the fire resistance integrity, and special detailing is necessary to close gaps and retain ability for independent movement. |
| | 2D-3 Gypsum wall board partitions | + | - | - | - | - | Light weight reduces effect of structural response in earthquakes. Although gypsum wallboard partitions can be constructed to have a fire rating, they can be easily damaged during fire events. Such partitions can be more easily damaged or penetrated during normal building use. |
| | 2D-4 Concrete block, hollow clay tile around exit ways and stairs | - | o | o | +/- | + | May create torsional response and/or stress concentrations in earthquakes, unless separated from structure, and if unreinforced, are prone to damage. Properly reinforced walls preserve evacuation routes in case of fire and blast. |

Table 10-1 Multihazard Design System Interactions (continued)

| | | Building System Protection Methods: Reinforcements and Conflicts | | | | | |
|---|---|---|---|---|---|---|---|
| | | **The Hazards** | | | | | |
| System ID | Existing Conditions or Proposed Protection Methods | Earthquakes | Flood | Wind | Security/Blast (FEMA 426) | Fire | Discussion Issues |
| **2E** | **Other Elements** | | | | | | |
| | 2E-1 Heavy ceramic/concrete tile roof | - | o | - | - | +/- | Heavy roofs undesirable in earthquakes; tiles may detach and fall. Provide good protection from fire spread, but can also cause collapse of fire-weakened structure. Dangerous in high winds unless very carefully attached. If a blast wave hits them they may become flying debris and dangerous to people outside the building. |
| | 2E-2 Parapets | +/- | o | + | - | + | Properly engineered parapet OK for seismic, but unbraced URM is very dangerous. May assist in reducing fire spread. |
| **3** | **Structural System** | | | | | | |
| | 3-1 Heavy Structure, RC masonry. Steel structure with masonry or concrete fireproofing | - | + | + | + | + | Increases seismic forces, requires sophisticated design. Generally beneficial against other hazards. |
| | 3-2 Light structure: steel/ wood | + | - | - | - | - | Decreases seismic forces but generally less effective against other hazards |
| | 3-3 URM load bearing walls | - | - | - | - | - | Poor performance against all hazards |
| | 3-4 RC or reinforced concrete block structural walls | + | + | + | + | + | Generally good performance against all hazards, provided correctly reinforced |
| | 3-5 Soft /weak first story (architectural/ structural design) | - | +/- | - | - | - | Very poor earthquake performance and vulnerable to blast. Generally undesirable for flood and wind. Elevated first floor is beneficial for flood, if well constructed and not in seismic zone. |
| | 3-6 Indirect load path | - | o | - | - | - | Undesirable for highly stressed structures, and fire-weakened structure is more prone to collapse. Not critical for floods. |

DESIGN FOR EXTREME HAZARDS

Table 10-1 Multihazard Design System Interactions (continued)

| | | The Hazards | | | | | |
|---|---|---|---|---|---|---|---|
| **Building System Protection Methods: Reinforcements and Conflicts** | | | | | | | |
| System ID | Existing Conditions or Proposed Protection Methods | Earthquakes | Flood | Wind | Security/Blast (FEMA 426) | Fire | Discussion Issues |
| | 3-7 Discontinuities in horizontal and vertical structure | - | o | - | - | - | Undesirable for highly stressed structures, causes stress concentrations, and fire-weakened structure is more prone to collapse. Not critical for floods. |
| | 3-8 Seismic separations in structure | + | o | o | o | - | Simplifies seismic response, possible paths for toxic gases in fires. |
| | 3-9 Ductile detailing of steel and RC structure and connections | + | o | + | + | o | Provides tougher structure that is more resistant to collapse. Not significant for fire. |
| | 3-10 Design certain elements for uplift forces | + | o | + | + | o | Necessary for wind, may assist in resisting blast or seismic forces. |
| | 3-11 RC, reinforced concrete blocks around exit ways and stairs. | - | o | o | -/+ | + | May create torsional structural response and/or stress concentration in earthquakes or blast. May preserve evacuation routes in the event of fires or blast. |
| **4** | **Building Envelope** | | | | | | |
| **4A** | **Wall cladding** | | | | | | |
| | 4A-1 Masonry veneer on exterior walls | - | - | - | - | o | In earthquakes, material may detach and cause injury. In winds and blast, may detach and become flying debris hazards. Flood forces can separate veneer from walls. |
| | 4A-2 Lightweight insulated cladding | + | o | o | - | o | Lightweight reduces structural response modification, may be less resistant to blast. |
| | 4A-3 Precast cladding panels | - | o | + | + | o | Require special detailing for earthquake. |
| **4B** | **Glazing** | | | | | | |
| | 4B-1 Metal/glass curtain wall | + | o | - | - | - | Light weight reduces earthquake forces, and if properly detailed and installed, performance is good. Fire can spread upward behind curtain wall if not properly fire-stopped. Not blast resistant without special glass and detailing. Vulnerable to high winds. |

Table 10-1 Multihazard Design System Interactions (continued)

| | | The Hazards | | | | | |
|---|---|---|---|---|---|---|---|
| System ID | Existing Conditions or Proposed Protection Methods | Earthquakes | Flood | Wind | Security/Blast (FEMA 426) | Fire | Discussion Issues |
| | 4B-2 Impact resistant glazing | o | o | o | + | - | Can cause problems during fire suppression operations, limiting smoke ventilation and access. Not significant for earthquake, flood or wind. |
| **5** | **Utilities** | | | | | | |
| | 5-1 Braced and well supported | + | o | + | + | + | Essential for earthquake, beneficial for wind, blast and fire. |
| **6** | **Mechanical** | | | | | | |
| | 6-1 System components braced and well supported | + | o | + | + | + | Essential for earthquake, beneficial for wind, blast and fire |
| **7** | **Plumbing and gas piping** | | | | | | |
| | 7-1 System components braced and well supported | + | o | + | + | + | Essential for earthquake, beneficial for wind, blast and fire |
| **8** | **Electrical and communications equipment** | | | | | | |
| | 8-1 System components braced and well supported | + | o | + | + | + | Essential for earthquake, beneficial for wind, blast and fire |
| **9** | **Fire suppression system and alarm** | | | | | | |
| | 9-1 System components braced and well supported | + | o | + | + | + | Essential for earthquake, beneficial for wind, blast and fire |

Building System Protection Methods: Reinforcements and Conflicts

CPSIA information can be obtained
at www.ICGtesting.com
Printed in the USA
LVOW05s2036081017
551670LV00008B/572/P